BIOCHEMISTRY RESEARCH TRENDS

CELLULASE

PRODUCTION, APPLICATIONS AND HEALTH BENEFITS

BIOCHEMISTRY RESEARCH TRENDS

Additional books in this series can be found on Nova's website under the Series tab.

Additional e-books in this series can be found on Nova's website under the eBooks tab.

BIOCHEMISTRY RESEARCH TRENDS

CELLULASE

PRODUCTION, APPLICATIONS AND HEALTH BENEFITS

MARGUERITE STEVENSON
EDITOR

NOVINKA

New York

NOTICE TO THE READER

Additional color graphics may be available in the e-book version of this book.

Library of Congress Cataloging-in-Publication Data

Names: Stevenson, Marguerite, editor.
Title: Cellulase : production, applications and health benefits / editor, Marguerite Stevenson.
Description: Hauppauge, New York : Nova Science Publisher's, Inc., [2016] |
Series: Biochemistry research trends | Includes bibliographical references and index.
Identifiers: LCCN 2016029814 (print) | LCCN 2016039822 (ebook) | ISBN 9781634857888 (softcover) | ISBN 9781634858021 (ebook) | ISBN 9781634858021
Subjects: LCSH: Cellulase.
Classification: LCC QP609.C37 C44 2016 (print) | LCC QP609.C37 (ebook) | DDC 572/.756--dc23
LC record available at https://lccn.loc.gov/2016029814

Published by Nova Science Publishers, Inc. † New York

Contents

Preface		**vii**
Chapter 1	Cellulase Producing Microorganisms: Diversity, Enzyme Production and Benefits *Saowapar Khianngam and Somboon Tanasupawat*	**1**
Chapter 2	Microbial Cellulase: Production and Applications in Enzymatic Treatment of Biosolids *A. A. Al-Gheethi, A. N. Efaq, R. M. S. R. Mohamed, J. D. Bala and M. K. Amir Hashim*	**45**
Chapter 3	Cellulase Properties, Market and Challenges of Lignocellulosic Material Hydrolysis *María García-Aparicio, Michel Brienzo and Johann F. Görgens*	**63**
Chapter 4	Cellulase Challenges of Lignocellulosic Material Hydrolysis *Michel Brienzo, María García-Aparicio, Luvuyo Tyhoda, Celso Sant'Anna and Johann Görgens*	**95**
Related Nova Publications		**121**
Index		**123**

PREFACE

Enzymes such as cellulases have been applied as tools for research and development and in almost every industrial sector including agriculture, food, beverages and animal feed, pulp and paper, textile. Cellulases is a general term that encompasses multiple enzymes that work in synergism for cellulose degradation. This book examines the production, applications and health benefits of cellulase. Chapter One describes the isolation, screening, diversity and cellulase production of microorganisms including their applications. Chapter Two discusses the factors affecting production of cellulase enzyme by microorganisms, the potential of indigenous microorganism for the bio-augmentation process of biosolids and occurrence of catabolite repression phenomenon of cellulase production during the enzymatic degradation of biosolids. Chapter Three gives an overview on cellulase microorganism producers, cellulolytic components and mechanism of action, with a focus on fungal enzymes of *Trichoderma reesei*, since they are predominant on the commercial cocktails used for 2G ethanol production. Chapter Four addresses substrate factors related to enzymatic hydrolysis of cellulose and lignocellulosic materials.

Chapter 1 – Cellulases are glycoside hydrolases involved in the depolymerization of cellulosic materials and have wide industrial applications. Cellulose, an abundant natural biopolymer and the most abundant of agricultural wastes, is a renewable resource with great potential for bioconversion to value-added bioproducts. It can be degraded by cellulase produced by microorganisms. Microbial isolation based on their halo zone formation on Congo red agar medium and efficient cellulase screening techniques and assays are described. Cellulase producing microorganisms are isolated from various sources and environments including soils, organic fertilizers, paper industry wastes, municipal wastes, sugarcane farms, gardens, wooded regions and marine sources. Diverse cellulase producing microbial genera including *Acidothermus*, *Achromobacter*, *Acinetobacter*, *Aeromonas*, *Aneurinibacillus*, *Anoxybacillus*, *Bacillus*, *Brucella*, *Cellulosimicrobium*, *Clostridium*, *Cellulomonas*, *Enterobacter*, *Enhydrobacter*, *Geobacillus*, *Isoptericola*, *Klebsiella*, *Lysinibacillus*, *Mesorhizobium*, *Micrococcus*, *Morganella*, *Ochrobactrum*, *Paenibacillus*, *Providencia*, *Pseudomonas*, *Xanthomonas*, *Shigella*, *Sphingomonas*, *Vibrio*, *Yersinia* and *Streptomyces*, *Aspergillus*, *Acremonium*, *Candida*, *Chaetomium*, *Fusarium*, *Mucor*, *Myceliophthora*, *Penicillium*, *Schizophyllum*, *Stachybotrys*, *Sporotrichum*, *Talaromyces*, *Thermoascus*, *Thermomucor* and *Trichoderma* strains have been screened and isolated. Industrial production of cellulases is being done using submerged fermentation processes and modified solid state fermentation systems. Cellulases are in demand worldwide and are used inbiofuel production, laundry and detergent industries, textiles, breweries, food, agriculture and human health. Additionally, the animal feed industry uses cellulases to improve the digestibility of plant products. In this chapter, we described the isolation, screening, diversity and cellulase production of microorganisms including their applications.

Chapter 2 – One of the main objectives in the treatment of biosolids is to reduce the cellulolytic materials which represent 40 to 50% of the total dry weight. Cellulose degradation is initiated primarily by micro-organism (fungi, bacteria and protozoa) with the aid of extracellular enzymes. The cellulase(s) is a multiple enzyme system consisting of endoglucanase, exoglucanase and β-glucosidase which act as a synergistic effect for the degradation of complex cellulose to glucose. Cellulase enzyme has applied for several industrial activities including production of animal feed, and wine, formulation of detergents, juice clarification, as well as in the treatment of domestic and industrial wastes. The enzymatic treatment of biosolids is named white technology which has raised recently due to the efficiency and absence of toxic by-products. In this chapter the factors affecting production of cellulase enzyme by microorganisms, the potential of indigenous microorganisms for the bio-augmentation process of biosolids and occurrence of catabolite repression phenomenon of cellulase production during the enzymatic degradation of biosolids will be discussed.

Chapter 3 – The increasing research interest on cellulases have been mainly driven by the production of second generation (2G) biofuels such as ethanol. As such, it is predicted that cellulases experience the largest expansion among the biofuels enzymes. The lignocellulose biomass presents a complex organization making the carbohydrate fraction not easily accessible by cellulases and other auxiliary enzymes. Thus a pretreatment step is necessary to overcome the material recalcitrance, making the carbohydrates more accessible to enzymes action. Progress on pretreatment technologies, reduction of cellulase production cost, and production of more robust cellulases with increased specific activity, have led to considerable reduction of enzyme requirements for lignocellulose hydrolysis. Moreover, the enzymatic hydrolysis optimization based on feedstock and type and conditions of pretreatment applied, can further reduce the enzyme requirements for carbohydrate saccharification into fermentable sugars. Despite these advances, enzyme production and its use are still considered one of the main bottlenecks for commercialization of cellulosic ethanol. A better understanding of the factors influencing the enzymatic hydrolysis of lignocellulose is essential to improve the process. These factors can be grouped into substrate-related and enzyme-related. Enzyme-related factors include the source of enzyme, expression system, biochemical properties of the enzymes, degree of synergism among the cellulolytic components and accessory enzymes, as well as the reaction conditions. In this chapter we will give an overview on cellulase microorganism producers, cellulolytic components and mechanism of action, with a focus on fungal enzymes of *Trichoderma reesei,* since they are predominant on the commercial cocktails used for 2G ethanol production. Factors limiting the enzymatic hydrolysis of lignocellulosic materials and current research trends to counteract them will also be presented.

Chapter 4 – Cellulases comprise a group of glycoside hydrolytic enzymes that cleave β-1,4-glycosidic linkages of cellulose into glucose units. These enzymes have been applied industrially for more than three decades. They have found application in different industrial sectors namely textile, detergents, pulp and paper, food, beverages and the feed industry. Notwithstanding, research in cellulase enzymes and their mechanism of action has increased steadily owing to the interest in the production of biofuels such as ethanol from lignocellulose materials. The biological conversion of cellulosic material is a sustainable approach focused on waste reduction and valorization. However, cellulose in lignocellulosic materials is highly organized and requires that the biomass must be pretreated and enzymatically hydrolyzed to release the fermentable sugars in appreciable amounts. Cellulose is associated with hemicelluloses and lignin in the rigid plant cell wall structures, which are naturally resistant to microbial and enzymatic degradation. The enzymatic hydrolysis of cellulose faces technical challenges related to both the nature of the substrate and enzyme factors. Despite the additional pretreatment step, there is normally a fraction of cellulose that remains un-hydrolyzed after enzymatic hydrolysis. Cellulases also display reduced stability under industrial application conditions such as temperature, the presence of inhibitors and end-product inhibition. This chapter addresses substrate factors related to enzymatic hydrolysis of cellulose and lignocellulosic materials. The cellulose and lignocellulosic materials are reviewed based on their chemical composition and physicochemical properties. Furthermore, these properties are related to the challenge of cellulases hydrolysis of cellulose and cellulosic materials. The effective hydrolysis of cellulose would allow for economically feasible production of biofuels (ethanol, hydrogen), chemicals and polymers.

In: Cellulase
Editor: Marguerite Stevenson
ISBN: 978-1-63485-788-8

Chapter 1

CELLULASE PRODUCING MICROORGANISMS: DIVERSITY, ENZYME PRODUCTION AND BENEFITS

Saowapar Khiangam[1] and Somboon Tanasupawat[2,*]
[1]Faculty of Animal Sciences and Agricultural Technology,
Silpakorn University, Phetchaburi IT Campus,
Phetchaburi, Thailand
[2]Department of Biochemistry and Microbiology,
Faculty of Pharmaceutical Sciences, Chulalongkorn University,
Bangkok, Thailand

ABSTRACT

Cellulases are glycoside hydrolases involved in the depolymerization of cellulosic materials and have wide industrial applications. Cellulose, an abundant natural biopolymer and the most abundant of agricultural wastes, is a renewable resource with great potential for bioconversion to value-added bioproducts. It can be degraded by cellulase produced by microorganisms. Microbial isolation based on their halo zone formation on Congo red agar medium and efficient cellulase screening techniques and assays are described. Cellulase producing microorganisms are isolated from various sources and environments including soils, organic

* Corresponding author: SomboonTanasupawat, Tel: +66-2-2188376, Fax: +66-2-2545195, E-mail : Somboon.T@chula.ac.th.

fertilizers, paper industry wastes, municipal wastes, sugarcane farms, gardens, wooded regions and marine sources. Diverse cellulase producing microbial genera including *Acidothermus*, *Achromobacter*, *Acinetobacter*, *Aeromonas*, *Aneurinibacillus*, *Anoxybacillus*, *Bacillus*, *Brucella*, *Cellulosimicrobium*, *Clostridium*, *Cellulomonas*, *Enterobacter*, *Enhydrobacter*, *Geobacillus*, *Isoptericola*, *Klebsiella*, *Lysinibacillus*, *Mesorhizobium*, *Micrococcus*, *Morganella*, *Ochrobactrum*, *Paenibacillus*, *Providencia*, *Pseudomonas*, *Xanthomonas*, *Shigella*, *Sphingomonas*, *Vibrio*, *Yersinia* and *Streptomyces*, *Aspergillus*, *Acremonium*, *Candida*, *Chaetomium*, *Fusarium*, *Mucor*, *Myceliophthora*, *Penicillium*, *Schizophyllum*, *Stachybotrys*, *Sporotrichum*, *Talaromyces*, *Thermoascus*, *Thermomucor* and *Trichoderma* strains have been screened and isolated. Industrial production of cellulases is being done using submerged fermentation processes and modified solid state fermentation systems. Cellulases are in demand worldwide and are used inbiofuel production, laundry and detergent industries, textiles, breweries, food, agriculture and human health. Additionally, the animal feed industry uses cellulases to improve the digestibility of plant products. In this chapter, we described the isolation, screening, diversity and cellulase production of microorganisms including their applications.

1. INTRODUCTION

Agricultural plant residues are a large and renewable source of lignocellulosic biomass that chiefly unexploited and inexpensive. Plant biomass contains polymers of 5- and 6-carbon sugars and is primarily comprised of cellulose (a homologous polymer of glucose linked by β-1,4 glycosidic bonds). It can be hydrolyzed by a complex enzyme system referred to as cellulases (exoglucanase, endoglucanase and β-glucosidase, etc.). Biomass also contains lesser hemicelluloses (heterologous polymers of 5- and 6-carbon sugars consisting of pentoses D-xylose, D-arabinose and hexoses D-mannose, D-glucose, D-galactose with sugar acids) and least of all lignin (a complex aromatic polymer) (Sadhu and Maiti, 2013). Fungal and bacterial strains have been used for the production of cellulases, however the fungi have more cellulolytic potential than bacteria and are most suited for extracellulase production. *Trichoderma reesei*, the most extensively studied species, converts native and derived cellulose to glucose, while bacteria, *Cellulomonas* and *Cytophaga*, can utilize a variety of carbohydrates and cellulose (Sukumaran et al., 2005). Actinomycetes are the other species of known cellulase producers have potential for cellulose production and help considerably in recycling nutrients in the biosphere (Bhat, 2000; Jang and Chen, 2003).

Cellulases are used to hydrolyze pretreated cellulosic material into sugars, which can be fermented to bioethanol and biobased products on a large scale. Additionally, there are applications of cellulases in the textile industry for biopolishing of fabrics and producing the stonewashed look of denim, as well as in household laundry detergents for improving fabric softness and brightness (Sadhu and Maiti, 2013). They are also used in animal feeds for improving nutritional quality and digestibility, in processing of fruit juices and in baking, while de-inking of paper is yet another emerging application. A potential challenging area where cellulases would have a central role is the bioconversion of renewable cellulosic biomass into commodity chemicals (Lynd et al., 2005). Currently, enzyme preparations suitable for biotechnology from several commercial enzyme producers are being marketed. The present chapter highlights background information on cellulase producing microorganisms.

2. Isolation and Screening of Cellulase Producing Microorganisms

2.1. Bacteria

The isolation and screening of cellulase producing bacteria (CPB) are typically performed on plates containing crystalline cellulose or microcrystalline cellulose in agar. Kasana et al., (2008) isolated bacteria from soils collected in western Himalaya, Himachal Pradesh, India using ten-fold serial dilutions and a spread plate method. They cultivated the isolates on nutrient agar at 28°C for 48 h. Five microliters of overnight grown cultures were spot plated on CMC 1 agar (Table 1) and incubated at 28°C for 48 h. The plates were flooded with 1% hexadecyltrimethylammonium bromide (HBA) for 30 to 40 min. These cultures were spot plated on CMC 1 agar and were stained with 0.1% Congo red for 15 to 20 min, followed by destaining with 1 M NaCl for 15 to 20 min to obtain zones of hydrolysis. Additionally, CMC plates were flooded with Gram's iodine instead of the above mentioned reagents. Gram's iodine forms a bluish-black complex with cellulose, but not with hydrolyzed cellulose, giving a sharp and distinct zone around the cellulase-producing microbial colonies within 3 to 5 min.

Table 1. Sources and media for isolation and screening of cellulase producing bacteria

Sources	Isolation and screening medium	References
Soil	1.**CMC broth (CMCB)** (g/L: CMC 10.0, algae powder 10.0, peptone 10.0, $(NH_4)_2SO_4$ 2.4, K_2HPO_4 2.0 and $MgSO_4.7H_2O$ 0.3, pH 7.4) 2. **CMC agar (CMCA)** (CMCB+1.5% agar, pH 7.4)	Yin et al., 2010
Compost pile	**CMC agar** [Nitrilotriacetic acid (100.0 mg/L), $CaSO_4$ (60.0 mg/L), $MgSO_4$ (100.0 mg/L), KNO_3 (103.0 mg/L), $NaNO_3$ (689.0 mg/L), Na_2HPO_4 (111.0 mg/L), $MnSO_4$ (2.2 mg/L), $ZnSO_4$ (0.5 mg/L), $CuSO_4$ (0.016 mg/L), H_3BO_3 (mg/L), Na_2MoO_4 (0.025 mg/L), $CoCl_2$ (0.046 mg/L), $FeCl_3.6H_2O$ (0.28 mg/L), tryptone (1.0 g/L), yeast extract (1.0 g/L), CMC (10.0 g/L) and agar 25.0 g/L, pH 7.0]	Ibrahim and El-Diwany, 2007; Acharya et al., 2012
Paper mill sludges and organic fertilizers: black-coloured solid sludge and wood-like solid sludge	1. **R2A** (g/L: yeast extract 0.5, protease peptone 0.5, casamino acids 0.5, glucose 0.5, soluble starch 0.5, K_2HPO_4 0.3, $MgSO_4.7H_2O$ 0.5, sodium pyruvate 0.3 and agar 15.0) 2. **LB liquid** (g/L: peptone 10.0, yeast extract 5.0 and NaCl 5.0) 3.**Carboxymethyl cellulose** (CMC) agar (g/L: CMC 5.0, $NaNO_3$ 1.0, K_2HPO_4 1.0, KCl 1.0, $MgSO_4$ 0.5, yeast extract 0.5 and agar 15.0)	Maki et al., 2011
Soil	**CMC agar** (g/L: peptone 10.0, carboxymethylcellulose (CMC) 10.0, K_2HPO_4 2.0, agar 10.0, $MgSO_4.7H_2O$ 0.3, $(NH_4)_2SO_4$ 2.5 and gelatin 2.0, pH 7.0)	Irfan et al., 2012; Patagundi et al., 2014
Soil from mangrove forest	**Carboxymethylcellulose** (CMC) agar (g/L: yeast extract 2.0, KH_2PO_4 1.0, $MgSO_4$ 5.0, and carboxymethylcellulose 5.0, pH 7.0)	Apun et al., 2000; Behera et al., 2014
Soil	1. **Czapek's medium** (g/L: $NaNO_3$ 2.0, $MgSO_4.7H_2O$ 0.5, NaCl 0.5, $FeSO_4.7H_2O$ 0.01, KH_2PO_4 1.0, yeast extract 0.4, sugarcane bagasse pulp 5.0 (containing 80% water) and agar 15.0, pH 5.0) 2. **Mandels and Reese medium** (g/L: KH_2PO_4 2.0, $(NH_4)_2SO_4$ 1.4, $MgSO_4.7H_2O$ 0.3, $CaCl_2$ 0.3, yeast extract 0.4, $FeSO_4.7H_2O$ 0.005, $MnSO_4$ 0.0016, $ZnCl_2$ 0.0017, $CoCl_2$ 0.002, CMC-Na 5.0 and agar 15.0, pH 5.0)	Wolf, 1952; Mandels and Reese, 1957; Liang et al., 2014
Soil	1. **Nutrient broth** (NB) (g/L: beef extract 3.0, peptone 5.0 and NaCl 5.0, pH 7.0) 2. **CMC agar** (g/L: $NaNO_3$ 2.0, K_2HPO_4 1.0, $MgSO_4$ 0.5, KCl 0.5, CMC sodium salt 2.0, peptone 0.2 and agar 17.0, pH 7.0) 3. **CMC Congo red agar** (g/L: K_2HPO_4 0.5, $MgSO_4$ 0.25, CMC sodium salt 1.88, Congo red 0.2, agar 15.0, gelatin 2.0 and 10.0% soil extract, pH 7.0)	Kasana et al., 2008; Chantarasiri et al., 2015
Coffee exocarp	1. **Mineral salt medium**(MM) (g/L: $Na_2HPO_4.2H_2O$ 3.5, KH_2PO_4 1.0, $(NH_4)_2SO_4$ 0.5, $MgCl_2.6H_2O$ 0.1, $Ca(NO_3)_2.4H_2O$ 50.0 mg, vitamin B_{12} 1 mL and trace salt solution 1 mL (g/100 mL of trace salt solution: MoO_3 0.01, $ZnSO_4.5H_2O$ 0.07, $CuSO_4.5H_2O$ 0.005, H_3BO_3 0.01, $MnSO_4.5H_2O$ 0.01, $CoCl_2.6H_2O$ 0.01 and $NiSO_4.7H_2O$ 0.01, pH 7.2] 2. **Cellulose agar**(g/L: CMC 10.0, $(NH_4)_2SO_4$ 2.5, yeast extract 0.5, KH_2PO_4 2.7, Na_2HPO_4 5.3, NaCl 0.2, $MgSO_4.7H_2O$ 0.2 and $CaCl_2$ 0.05, pH 7.2-7.4)	Crawford, 1983; Fortnagel et al., 1990; Bui, 2014

Yin et al., (2010) isolated CPB from the soil in northern Taiwan using 10% of sample suspended in sterile phosphate-buffered saline (PBS, 10 mM potassium, 150 mM NaCl, pH 7.4). The diluted sample was inoculated and incubated in a 1.0% CMC broth at 30°C for 7 days. Ten percent of the resulting broth was inoculated into CMC broth (CMCB) and incubated at 30°C for 3 days. The resulting broth was then spread on a CMC agar plate (CMCA, CMCB + 1.5% agar) and incubated at 30°C for 3 days. The colonies with clear zones (indicating those with CMC hydrolyzing ability) on CMCA were selected and inoculated onto a fresh CMCA plate. After 3 days of incubation at 30°C, the colonies with the largest CMC hydrolyzing halo rings were picked and subjected to identification. The isolated strain was stored at -80°C in CMCB with 20% (v/v) glycerol for further studies. CMCB was employed for further production of cellulase under aerobic incubation at 30°C with 150 rpm shaking.

Acharya et al., (2012) isolated thermophilic cellulolytic bacteria from 13 different compost samples (temperature >50°C). The compost samples were air dried and incubated at 55°C for 1 week and to reduce the mesophilic and anaerobic isolates. An appropriate dilution of 1 g compost samples was then both direct spread plated on CMC agar medium or enriched in cellulose broth and plated (spread) on CMC agar and then incubated at 55°C for 2-4 weeks with shaking at 120 rpm (Ibrahim and El-Diwany, 2007). Congo red (1%) and 1M NaCl solutions were used to indicate cellulose hydrolysis. If the zones were not clear, 0.1 N HCl was added to make them clearer (Wood and Bhat, 1988; El-Sersy et al., 2010).

Cellulolytic bacteria were isolated from 1 g of each sample and suspended in 20 mL of sterile PBS, vortexed for 2 min at maximum speed and the suspension was diluted as a 10 x serial dilution in PBS. Two hundred microliters of each dilution in the series was spread onto the surface of R2A agar and the plates were incubated at 28°C for 24-72 h. Colonies were selected based on their morphology, size and colour (Maki et al., 2011). Screening for carboxymethyl cellulase activity was performed by cultivating the isolates in 10 mL of LB broth (Table 1) for 24-48 h, with agitation at 28°C. A *Cellulomonas xylanilytica* strain was used as a positive control and incubated for 5 days while *Escherichia coli* JM109 was a negative control and was incubated at 37°C for 18 h. In this method, 5 µL of each broth culture were singly dropped onto a plastic petri dish containing carboxymethylcellulose (CMC) agar (Table 1) and then incubated for 48 h at 28°C. The positive control was incubated for 60 h longer than the isolates and negative control due to its slow growth rate. After the allotted growth time, the CMC agar plates with the isolates and controls were stained with a Gram's iodine solution (2.0 g KI and 1.0 g I, per 300 mL ddH_2O) to visualize the cellulase activity (Maki et al., 2011).

Irfan et al., (2012) isolated cellulolytic bacteria from soil by using serial dilutions and a pour plate technique on carboxymethylcellulose (CMC) agar (Table 1) incubated at 30°C for 48 h. Pratima et al., (2012) isolated cellulose degrading microorganisms using an agar medium comprised ofon a per liter basis: 0.5 g KH_2PO_4, 0.25 g $MgSO_4$, 2.0 g cellulose, 15.0 g agar, and 2.0 g gelatin at pH 6.8-7.2 and using the medium reported by Crawford (1983). The cellulolytic bacteria from the soil sample was collected from a botanical garden at Karnatak University Campus, Karnataka, India and was isolated using a serial dilution agar plating method with cultivation on a carboxymethylcellulose medium (Table 1) and incubated for 48 hours at 30°C (Patagundi et al., 2014). Behera et al., (2014) isolated cellulose degrading bacteria from soil samples collected from mangrove forests at Jumbo, Kharnasi, Triveni, Nuagada, Atharabanki and a mangrove forest at the Indian Farmer Fertiliser Corporation (IFFCO). From each soil sample, several sub-samples were taken, homogenized in sterile milli Q water containing 0.85% NaCl (w/v) and serially diluted. The pour plate and spread plate techniques were done using carboxymethylcellulose (CMC) agar plates and incubated at 37°C for 24-48 h. The plates were flooded with a 0.1% Congo red solution and washed with 1 M NaCl to visualize the hydrolysis zones.

Liang et al., (2014) collected organic-rich soils from Maoer Mountain (Guilin City), Longgang (Chongzuo City), Dawang Ridge (Baise City), Huaping (Guilin City), Shankou Halodrymium (Beihai City) Natural Reserves, a starch factory in Fangchenggang City, a bagasse compost at the experimental farm of Guangxi University (Nanning City) in Guangxi Zhuang Autonomous Region, China, and Baima Snow Mountain Natural Reserve in Yunnan Province, China. The soil sample suspensions were inoculated on Czapek's medium containing sugarcane bagasse pulp and incubated at 28°C. Subsequently, single colonies were picked using an inoculating needle and inoculated onto Mandels and Reese medium containing sodium carboxymethyl cellulose (Table 1). After incubation at 28°C for 48 h, all the plates were stained with a 1% (w/v) Congo-red solution as described above. The degradation zones were visible around the bacterial colonies, showing that the strains could hydrolyze CMC and the modified Mandels medium (also called basal medium) used for CMCase production (Table 1). Chantarasiri et al., (2015) collected soil samples from Nong Chamrung Coastal Wetlands, and the Rayong Botanical Garden, Rayong Province, Thailand. Soil samples from the surface and at a depth of 15 cm were collected using sterile spatulas and stored in sterile zip-lock polyethylene bags. Carboxymethylcellulose (CMC) agar plates were used to screen for cellulase producing bacteria as described by Kasana et al., (2008). Five microlitres of a culture of each bacterial isolate grown overnight in nutrient broth was spot plated on CMC agar (Table 1) and incubated at 29.2°C for 48 h. Gram's iodine solution was used to flood the plates for 5 min. *B. methylotrophicus* RYC01101 and *E. coli* were used as positive and negative controls for the screening, respectively.

2.2. Actinomycetes

Soil samples collected at a depth of 10 cm below the selected tree canopy were pre-treated by drying in open air for 1 week before isolation. Two grams of each soil sample were mixed with 20 mL of sterile distilled water and shaken (250 rpm) at room temperature (±25°C). Two hundred microlitres of the soil suspension was spread onto Humic Acid (HV) agar plates (Hayakawa et al., 2004) and incubated for 7 days at ±30°C. Grown colonies were picked using sterile toothpicks and touched onto starch casein agar (SCA) (Table 2) and incubated for 7 days at ±30°C. Pure colonies of actinomycetes were obtained after several subcultures (Jeffrey and Azrizal, 2007). Rathnan and Ambili (2011) isolated actinomycete strains from decayed wastes, including the fruit waste, orange bagasse, banana and mango peels using a starch casein medium with incubation for 7 days at 28°C. The isolates were screened on carboxymethylcellulose (CMC) agar medium with 1% of cellulose and incubated at 28°C for 4 days. A hydrolysis zone was identified around the culture by treating the plates with Congo red and NaCl.

Bui (2014) isolated cellulolytic bacteria and actinomycetes from 15 coffee exocarp samples collected in Vietnam. A mineral salt medium (MM) was used for enrichment and filter paper (Whatman No. 1 with an area 70.541 cm^2) was used as the substrate for enrichment and growth (Fortnagel et al., 1990). For coffee exocarp samples, enrichment culture was started by mixing 10.0 g of the samples with 100 mL of distilled water and shaking for 1 h on a rotary shaker at 250 rpm. The sample (100 mL) was mixed with 200 mL of a mineral salt medium supplemented with filter paper and incubated aerobically at 30°C on a rotary shaker at 150 rpm (Fortnagel et al., 1990). Aliquots were transferred weekly from the culture to a fresh medium. Subcultures were streaked onto a nutrient medium, and incubated aerobically at 30°C. After incubation at 30°C, single colonies were transferred onto mineral salt agar media containing the above carbon source. Well grown pure colonies were picked and transferred onto a mineral salt agar medium, and were stored at 4°C until use insubsequent experiments.

Table 2. Sources and media for isolation and screening of cellulase producing actinomycetes

Sources	Isolation and screening medium	References
Soil	1. **Humic Acid (HV) agar** [g/L: humic acid 1.0, Na_2HPO_4 0.5, KCl 1.7, $MgSO_4.7H_2O$ 0.05, $FeSO_4.7H_2O$ 0.01, $CaCl_2$ 1.0, B-vitamins (0.5 mg each of thiamine-HCl, riboflavin, niacin, pyridoxin, capantothenate, inositol, *p*-aminobenzoic acid, and 0.25 mg of biotin) and agar 18.0 g, pH 7.4] 2. **Starch Casein Agar** (SCA) [g/L: soluble starch 10.0, casein hydrolysate 0.3, KNO_3 2.0, NaCl 2.0, KH_2PO_4 2.0, $MgSO_4.7H_2O$ 0.05, $CaCO_3$ 0.02, $FeSO_4.7H_2O$ 0.01, agar 18.0 and cycloheximide (100 µg/mL)] 3. **Cellulases activity using substrate agar** (g/L: peptone 1.0, yeast extract 1.0, $MgSO_4.7H_2O$ 0.5, K_2HPO_4 0.5, $(NH_4)_2HPO_4$ 1.0, Azo-CM-Cellulose (Megazyme) 1.0 and agar 15.0)	Hayakawa et al., 2004; Jeffrey et al., 2007
Coffee exocarp	1.**Mineral salt medium (MM) (g/L:** $Na_2HPO_4.2H_2O$ 3.5, KH_2PO_4 1.0, $(NH_4)_2SO_4$ 0.5, $MgCl_2.6H_2O$ 0.1, $Ca(NO_3)_2.4H_2O$ 50 mg, vitamin B_{12} 1 mL and trace salt solution 1 mL (g/100 mL of trace salt solution: MoO_3 0.01, $ZnSO_4.5H_2O$ 0.07, $CuSO_4.5H_2O$ 0.005, H_3BO_3 0.01, $MnSO_4.5H_2O$ 0.01, $CoCl_2.6H_2O$ 0.01 and $NiSO_4.7H_2O$ 0.01, pH 7.2] 2. **Cellulose agar** (g/L: CMC 10.0, $(NH_4)_2SO_4$ 2.5, yeast extract 0.5, KH_2PO_4 2.7, Na_2HPO_4 5.3, NaCl 0.2, $MgSO_4.7H_2O$ 0.2 and $CaCl_2$ 0.05, pH 7.2-7.4)	Crawford, 1983; Fortnagel et al., 1990; Bui, 2014
Soil	1. **Yeast extract-malt extract (YM agar)** (g/L: yeast extract 4.0, malt extract 10.0, dextrose 4.0 and agar 20.0, pH 7.3) 2. **Starch Casein (SC agar)** (g/L: soluble starch 10.0, K_2HPO_4 2.0, KNO_3 2.0, casein 0.3, $MgSO_4.7H_2O$ 0.05, $CaCO_3$ 0.02, $FeSO_4.7H_2O$ 0.01 and agar 15.0, pH: 7.0) 3. **Arginine Glycerol (AG agar)** (g/L: arginine monohydrochloride 1.0, glycerol 12.5, K_2HPO_4 1.0, NaCl 1.0, $MgSO_4.7H_20$ 0.5, $Fe_2(S0_4)_3.6H_20$ 0.01, $CuSO_4.5H_20$ 0.001; $ZnSO_4.7H_20$ 0.001, $MnSO_4.H_20$ 0.001 and agar 15.0, pH 6.9-7.1) 4. **Organic Agar Gause 2 (OMG 2)** (g/L: soluble starch 20.0, KNO_3 1.0, K_2HPO_4 0.5, $MgSO_4.7H_2O$ 0.5, $FeSO_4.7H_2O$ 0.01 and agar 15.0, pH 7.2-7.4)	El-Nakeeb and Lechevalier, 1963; Kuester and Williams, 1964; Shirling and Gottlieb, 1966; Gause et al., 1983; Das et al., 2014

Sources	Isolation and screening medium	References
	5.**Glycerol Asparagine (GA agar)** [g/L: L-asparagine 1.0, glycerol 10.0, K_2HPO_4 1.0, 1 mL of trace salts solution (g/L: $FeSO_4.7H_2O$ 1.0, $MnCl_2.4H_2O$ 1.0, $ZnSO_4.7H_2O$ 1.0) and agar 20.0, pH 7.0-7.4] 6. **Basal agar medium containing** [g/L: $(NH_4)_2SO_4$ 2.64, KH_2PO_4 2.38, $K_2HPO_4.3H_2O$ 5.65, $MgSO_4.7H_2O$ 1.0, 1 mL of trace salt solution (composition g/L, $CuSO_4.5H_2O$ 6.43, $FeSO_4.7H_2O$ 1.1, $MnCl_2.4H_2O$ 7.9, $ZnSO_4.7H_2O$ 1.5) and agar 15.0 (pH 6.8-7.0) supplemented with 2% cellulose]	
Soil	1. **Starch Casein Agar** (g/L: soluble starch 10.0, casein 0.3, K_2HPO_4 2.0, $CaCO_3$ 0.02, $FeSO_4.7H_2O$ 0.01, KNO_3 2.0, $MgSO_4.7H_2O$ 0.05, NaCl 2.0 and agar 18.0, pH 7.0) 2. **CMC Agar** (g/L: carboxymethylcellulose 0.5, $NaNO_3$ 0.1, K_2HPO_4 0.1, $MgSO_4$ 0.05, yeast extract 0.05 and agar 15.0, pH 7.0)	Dubey and Maheshwari, 2004; Kasana et al., 2008; Prasad et al., 2014
Sediment, plant decomposed soil	1. **ISP-2 Medium** (g/L: yeast extract 4.0, malt extract powder 10.0, glucose 4.0 and agar 18.0) 2. **Carboxymethyl cellulose in a basal salt media** [g/L: yeast extract 1.0, CMC 5.0, agar 10.0 and mineral salts 12.5 mL (g/L: $(NH_4)_2SO_4$ 2.0, KH_2PO_4 4.0, Na_2HPO_4 6.0, $FeSO_4.7H_2O$ 1.0 mg, $MgSO_4$ 200.0 mg, $CaCl_2$ 1.0 mg, H_3BO_3 10 μg, $MnSO_4$ 10 μg, $ZnSO_4$ 70 μg, $CuSO_4$ 50 μg, MoO_3 10 ug), at pH 7.0]	Zucker and Hankin, 1970; Hankin and Anagnostakis, 1977; Mohanta, 2014
Persicaria glabra roots rhizosphere soil	1. **Starch casein broth** (g/L: K_2HPO_4 0.5, $MgSO_4.7H_2O$ 0.5, soluble starch 10.0 and casein 1.0, pH 7.0) 2. **Carboxymethyl cellulose (CMC) agar** (g/L: KH_2PO_4 1.0, $MgSO_4$ $.7H_2O$ 0.5, NaCl 0.5, $FeSO_4.7H_2O$ 0.01, $MnSO_4.H_2O$ 0.01, NH_4NO_3 0.3, CMC 10.0 and Agar 12.0)	Singh et al., 2015

Das et al., (2014) isolated actinomycetes from soil samples collected from diverse ecological habitats using a spread technique on specific media including Yeast Extract-Malt Extract (YM agar), Starch Casein (SC agar), Arginine Glycerol (AG agar), Organic Agar Gause 2 (OMG 2) and Glycerol Asparagine (GA agar) (Table 2). The isolated actinomycetes were then screened for their cellulase activities using substrate agar [1.0 g peptone, 1.0 g yeast extract, 0.5 g $MgSO_4.7H_2O$, 0.5 g K_2HPO_4, 1.0 g $(NH_4)_2HPO_4$, 1.0 g Azo-CM-Cellulose (Megazyme), 15.0 g agar; and 1,000 mL distilled water]. Halo zone formation after 3 days of incubation indicated a positive reaction.

Prasad et al., (2014) collected soil samples from a depth of 3-4 inches below an agricultural field, municipal dumps and domestic wastes in the Patna region of India. The samples were kept in polythene bags and brought to the laboratory for analysis. Starch Casein Agar was used for isolation and CMC agar was used for screening (Table 2). Cellulose-degrading actinomycetes were isolated from sediments and plant decomposed soils collected from the Bhitarkanika National Park by a dilution method using ISP-2 medium. The plates were incubated for 7 days at 28°C. They were screened for their cellulase ability in a plate assay method using 1% carboxymethyl cellulose in a basal salt medium (Mohanta, 2014). Singh et al., (2015) performed a preliminary screening using starch casein broth (100 mL for each strain) for inoculating the strains and kept them for 3 days of incubation in an orbital shaker incubator at 28°C and 120 rpm. These culture broths were diluted and used as an inoculum to obtain isolated colonies. Carboxymethyl cellulose (CMC) agar plates were inoculated with 100 μL of culture broth by spread plate technique and incubated at 37°C for 3 days to allow secretion of cellulase. Congo red dye and 1 M NaCl were used to determine the cellulase activity. To increase the contrast of halos, the plates were treated with 5 mL of 1M HCl.

2.3. Fungi

Cellulolytic fungi were isolated from waste paper, waste paper after primary re-sort, waste paper after deep resort (paper pulp), pulp after clarification, primary effluent mud, waste water a from biological treatment process, waste water from the sediment of a retention basin six months and again five years after deposit in a waste water retention basin at the "Klaipėdoskartonas" cardboard factory in Lithuania (Pečiulytė, 2007). A medium containing a Mandels' mineral salt solution (Mandels et al., 1974) added to 17.5 g/L agar, 5.0 g/L phosphoric acid-swollen cellulose and 0.5% L-sorbose as a colony restrictor and inducer of cellulase production (Wang et al., 1995) was used for the isolation. The cellulose substrate used in this clearing test was prepared according to the procedure of Stewart et al., (1982). The plates were seeded with 100 μL of a stock culture suspension (10 plates for each sample) and incubated at 25°C for 4-7 days, and then further incubated for 24 h at 50°C, after which clear zones could be observed only around colonies of active cellulase producing fungal strains. Cellulase activities of the more active fungal isolates were determined using a carboxymethylcellulase activity assay (CMCase). A basal medium containing (g/L): 10 g CMC, 6.5 g $NaNO_3$, 6.5 g K_2HPO_4, 0.3 g yeast extract, 6.5 g KCl, 3.0 g $MgSO_4.7H_2O$, 0.65 g glucose and 17.5 g agar, was used for plate screening. Agar blocks (5 mm in diameter) from one-week old colonies grown on PDA plates were cut and inoculated in the center of the plates. The plates were incubated at 25°C for three to five days, and then for 18 h at 50°C in the dark. Cellulolytic strains were selected on the basis of the diameter of the hydrolysis zones surrounding the colonies. For observation, the plates were stained with 1% Congo red dye (15 min), followed by destaining with a 1 M NaCl solution for 20 min. Cellulase activity on carboxymethylcellulose (CMC, sodium salt medium viscosity, SIGMA) agar was used to determinean Index of Relative Enzyme Activity (ICMC) asa clear zone ratio, i.e., a ratio of clear zone diameter/colony diameter (Teather and Wood, 1982; Bradner et al., 1999).

Khokhar et al., (2012) isolated cellulolytic filamentous fungi from soil, industrial effluents, fruit seeds, vegetables, bread and wood using malt extract agar (MEA), potato dextrose agar (PDA) and Czapek's agar (CA). Cellulase activities of the fungal isolates were determined using a plate screening medium (PSM) (Table 2). Agar blocks (8 mm in diameter) from one-week old fungal colonies grown on MEA plates were cut and inoculated in the centre of basal media plates. The plates were incubated at 25± 2°C for seven days. Cellulolytic fungal species were selected on the basis of the diameter of the hydrolysis zones surrounding the colonies. The plates were stained with 1% Congo red dye (30 min), followed by destaining with a 1 M NaCl solution for 20 min. Clear zones could be observed only around colonies of the active cellulytic fungal strains. Cellulase activity on carboxymethyl agar was recorded as an Index of Relative Enzyme Activity (ICMC) as described above (Teather and Wood, 1982; Bradner et al., 1999; Pečiulytė, 2007). Growth of fungal isolates on MEA was used as a control. A growth simulation/inhibition index was computed as the ratio of colony diameter on pectin agar/colony diameter on control agar.

Damaso et al., (2012) isolated cellulolytic fungi from soils, sugarcane bagasse, and plants by serial dilution on a PDA medium (Potato Dextrose Agar) with the addition of chloramphenicol. The isolations were performed in Erlenmeyer flasks by re-suspension of five grams of the sample in 45 mL of a peptone solution (0.1%) containing 0.1% Tween 80, followed by 30 min shaking at 150 rpm (Gomes et al., 2001; Markovina et al., 2005). Species identification was done on selective media for each group of fungi and was based on morphological studies.

Table 3. Sources and media for isolation and screening of cellulase producing fungi

Sources	Isolation and screening medium	References
Waste paper, mud, waste water	1.**Mendel's mineral salt** (g/L: urea 0.3, $(NH_4)_2SO_4$ 1.4, KH_2PO_4 2.0, $CaCl_2$ 0.3, $MgSO_4$ 0.3, yeast extract 0.25 and proteose peptone 0.75, phosphoric acid-swollen cellulose 5.0, L-sorbose 5.0 and agar 17.5, pH 4.0-6.0) 2. **Carboxymethyl cellulase basal medium** (g/L: CMC 10.0, $NaNO_3$ 6.5, K_2HPO_4 6.5, yeast extract 0.3, KCl 6.5, $MgSO_4.7H_2O$ 3.0, glucose 0.65 and agar 17.5)	Mandels et al., 1974; Wang et al., 1995; Pečiulytė, 2007
Soil	1. **Potato dextrose agar** (g/L: potatoes 200.0, dextrose 20.0 and agar 17.0) 2. **CMC agar** (g/L : $NaNO_3$ 2.0, K_2HPO_4 1.0, $MgSO_4$ 0.5, KCl 0.5, carboxymethylcellulose (CMC) sodium salt 2.0, peptone 0.2 and agar 17.0) 3. **CMC Congo red agar** (g/L: K_2HPO_4 0.5, $MgSO_4$ 0.25, CMC sodium salt 1.88, Congo red 0.2, agar 15.0, gelatin 2.0 and 10% soil extract)	Kasana et al., 2008
Soil, industrial effluents, seeds fruits, vegetables, bread and wood	1. **Malt extract agar (MEA)** 2. **Potato dextrose agar (PDA)** 3. **Czapek's agar (CA)** 4. **Plate screening medium (PSM) contained Mendel's mineral salt** (g/L: Urea 0.3, $(NH_4)_2SO_4$ 1.4, KH_2PO_4 2.0, $CaCl_2$ 0.3, $MgSO_4$ 0.3, yeast extract 0.25, proteose peptone 0.75, carboxymethyl cellulose (CMC) 10.0 and agar 17.5)	Mandels et al., 1974; Khokhar et al., 2012
Soil	**Czapek-Dox medium** (g/L: sucrose 30.0, $NaNO_3$ 2.0, K_2HPO_4 1.0, $MgSO_4$ 0.05, KCl 0.5, $FeSO_4$ 0.01, Carboxymethyl cellulose 10.0 and agar 20.0, pH 5.0)	Sri Lakshmi and Narasimha, 2012
Areca nut husk waste	1. **Martin Rose Bengal agar** (g/L: papaic digest of soyabean meal 5.0, dextrose 10.0, KH_2PO_4 1.0, $MgSO_4$ 0.5, rose Bengal 0.05 and agar 15.0, pH 7.2) 2. **Potato dextrose agar plates supplemented with 2% Carboxymethyl cellulose sodium salt** (CMC) and tetracycline (10-25 μg/mL), pH 7.0	Naveenkumar and Thippeswamy, 2013
Sources	**Isolation and screening medium**	**References**
Forest wastes like degrading wood, leaf litter and soil	1. **Potato dextrose agar (PDA)** 2. **Czapeck-Dox medium supplemented with Carboxymethyl cellulose (CMC)** (g/L: CMC 10.0, sucrose 30.0, $NaNO_3$ 2.0, K_2HPO_4 1.0, $MgSO_4$ 0.5, KCl 0.5, $FeSO_4$ 0.01 and agar 15.0, pH 7.3)	Kumar et al., 2014
Soil	**Berg's media** (g/L: $NaNO_3$ 2.0, $MgSO_4$ 0.5, K_2HPO_4 0.005, $FeSO_4$ 0.01, $CaCl_2$ 0.02, $MnSO_4$ 0.002, CMC 5.0 and agar 15.0, supplemented with Streptomycin)	Gupta et al., 2015

Cellulose degradation ability was evaluated on a basic medium (modified by Couri and Farias, 1995) with carboxymethylcellulose, sugarcane bagasse or "avicel" as the carbon source. Each isolated fungi was inoculated in the basic medium, distributed on Petri dishes and incubated at 25 or 32°C.Thereafter 10 mL of Congo red buffer (0.1 M Tris-HCl, pH 8.0) was added. After 15 min, the solution was discarded and the cultures were washed with 5.0 mL 0.5 M NaCl in the same buffer, and the hydrolysis halos were measured (Teather and Wood, 1982). Selection of filamentous fungi that were good producers of cellulolytic enzymes was carried out as follows: 1) measurement of the diameter of the cultivated colonies on the basic medium with carboxymethylcellulose as the carbon source, and 2) estimation of the enzymatic index with a Congo red stain solution. At the places where the carboxymethylcellulose substrate was hydrolyzed, a translucent halo was observed and the enzymatic index was evaluated as the ratio of the diameter of the halo to the diameter of the colony. Strains presenting indices near or greater than 1 (one) were selected.Lastly, enzymes were producedin a semi-solid fermentation using the strains selected from stages 1 and 2 (modified by Damaso et al., 2008).

Sri Lakshmi and Narasimha (2012) screened for cellulolytic fungi using Czapek-Dox medium (Table 3) (Teather and Wood, 1982). Naveenkumar and Thippeswamy (2013) isolated fungi using potato dextrose agar plates supplemented with a 2% carboxymethylcellulose sodium salt (CMC) (HiMedia) as a carbon source withtetracycline (10-25 μg/mL) to control bacterial contamination (pH 7.0), incubated at 28 ± 1°C for 5-7 days. After incubation, fungal species were purified and subcultured on carboxymethyl cellulose agar plates and used for the subsequent purpose (Gautam et al., 2010; Banakar et al., 2012). Secondary screening was done for the selection of more potent colonies for the production of extracellular cellulase by culturing on carboxymethyl cellulose agar plates. Kumar et al., (2014) isolated fungi from forest wastes such as decaying wood, leaf litter and soil collected from the forest near the Forest Research Institute, Dehradun, India. Cellulase enzyme activities ofisolated fungal strains were primarily determined by culturing on Czapeck-Dox medium supplemented with 1% carboxymethyl cellulose (CMC). After incubation at 28 ± 1°C for 5-7 days, plates were flooded with a Congo red solution (1 mg/mL in distilled water) for 20 min. The dye was decanted and the plates flooded with 5 M NaCl for 20-30 min, after which this solution was decanted. Carboxymethylcellulase (CMCase) producing colonies were surrounded by a pale orange to clear halo against the background (Teather and Wood, 1982; Banakar et al., 2012). Gupta et al., (2015) isolated cellulose degrading fungi from a soil sample using serial dilution method. 0.1 mL of each dilution was pipetted and spread onto Berg's media (Table 3) supplemented with streptomycin and incubated at 28°C for 3-7 days.

3. Diversity of Cellulase Producing Microorganisms

3.1. Bacteria

The distribution of cellulase producing microorganisms from various sources is shown in Table 4. *Clostridium cellulovorans*, a cellulolytic bacterium(Sleat et al., 1984) and *Acidothermus cellulolyticus* 11B from an acidic hot springs at Yellowstone National Park, Wyo, were proposed (Mohagheghi et al., 1986) as novel species exhibiting cellulolytic activity. Thermophilic cellulase producing bacteria, *Anoxybacillus flavithermus* EFP1, *Geobacillus thermodenitrificans* EFP2 and *Geobacillus stearothermophilus* EFP3 were isolated from an Egyptian hot spring by enrichment of the water and soil samples (Ibrahim and El-Diwany, 2007). A *Cellulosimicrobium cellulans*strain used bagasse, a coproduct of the sugar industry, as the sole carbon source for aliquid-state fermentation to produce hydrolytic enzymes (cellulases and xylanases) (Song and Wei, 2010). *Paenibacillus cellulositrophicus*, a novel cellulolytic bacterium from a Thai soil sample was isolated (Akaracharanya et al., 2009). *Cellulomonas*sp. YJ5 produced a maximal cellulase activity after 48 h of incubation at 30°C and thiscellulase was purified using Sephacryl S-100 chromatography. Substrate specificity indicated that it was an endo-beta-1,4-glucanase (Yin et al., 2010). *Paenibacillus* sp. P118 isolated from the gut of a tropical fish, *Parotocinclus maculicauda*, was a potential strain for the production of industrially important products from cellulosic biomass (de Castro et al., 2011).

An *Aeromonas* strain with cellulase-producing capabilities was isolated from the intestines of an indigenous fish species, Herbivorous Grass Carp (*Ctenopharyngodon idella*). It showed a powerful capability to digest aquatic weeds (Jiang et al., 2011). *Shigella flexneri* G3 displayed effective production of glucose, cellobiose and other oligosaccharides from cellulose (Avicel PH-101) under optimal conditions (40°C and pH 6.5) (Wang et al., 2011). *Aneurinibacillus thermoaerophilus* WBS2 isolated from a hot spring in India, produced maximalcellulase activity at pH 9.0 when growing in shaken culture

(150 rpm) at 65°C (Acharya and Chaudhary, 2012a). *Klebsiella variicola* RBEB3 produced a cellulase enzyme with a 45 kDa molecular weight as determined by an SDS PAGE method (Gopinath et al., 2012). *Cellulomonas* sp. ASN2 isolated from soil produced cellulase in 250 mL Erlenmeyer flasks using potato waste as a substrate in submerged fermentation (Irfan et al., 2012; Safdar et al., 2013). *Paracoccus pantotrophus* FMR19 isolated from goat rumen fluid produced maximalcellulaseactivity at pH 4.0 with 3600 units/mg and 1200 units/mg at 40°C (Faridha et al., 2013). *Mesorhizobium* sp. isolated from a marine source produced maximalcellulase activity using Eucalyptus as a substrate followed by rice straw (Prasad and Sethi, 2013). *Micrococcus* spp., *Bacillus*spp., *Pseudomonas* spp., *Xanthomonas* spp. and*Brucella*spp., from mangrove soil of the Mahanadi River Delta, Odisha, India exhibited maximal carboxymethylcellulose hydrolysis (HC value) which ranged from 1.18 to 2.5 cm (Behera et al., 2014).

Bacillus subtilis BY-2 isolated from a Tibetan pig's intestine produced cellulase using 1% corn flour as carbon source and 2% peptone and yeast powder as nitrogen sources (Yang et al., 2014). The cellulase enzyme from *Acinetobacter* sp. KKU44 is heat-tolerant (Siripornadulsil et al., 2014).*Bacillus* spp., *Paenibacillus* sp. and *Lysinibacillus* sp., from oil palm meal samples collected in Phetchaburi, Prachuap Khiri Khan and Pattani Provinces, Thailand, exhibited cellulase activity that ranged from 0.039±0.002 to 0.233±0.005 IU/mL (Khianngam et al., 2014).*Paenibacillus terrae* ME27-1 produced carboxymethylcellulase (CMCase) at its optimal pH 5.5 and at 50°C (Liang et al., 2014). Three cellulase producing bacteria from soil sampleswere collected from the Botanical Garden of Karnatak University Campus, Karnataka, India. *Bacillus cereus* showed a maximal enzyme activity (0.440 IU/mL/min and 0.410 IU/mL/min), followed by *B. subtilis* (0.357 IU/mL/min) and *B. thuringiensis* (0.334 IU/mL/min) (Patagundi et al., 2014).

B. cereus strain BR0302 from Thai coastal wetland soil exhibited its highest hydrolysis activity on carboxymethylcellulose agar plates (Chantarasiri et al., 2015). *Bacillus* sp. and *Ochrobactrum anthropi* (YZ1) isolated from soils of the Guntur District, A.P. showed equal maximal yields of cellulase enzymes (Duza and Mastan, 2015). *B. vallismortis* RG-07 showed maximalcellulase production from sugarcane baggase (4,105 U/mL) used as an agro-waste carbon source (Gaur and Tiwari, 2015). *Isoptericola variabilis* sp. DAH9 isolated from the hot springs of Dehloran, Ilam Province, Iran, produced a thermostable endoglucanase on low-priced carbon sources such as rice barn and commercial CMC (Azizi et al., 2015). *Enhydrobacter* sp. ACCA2 produced endoglucanase (CMCase) (Premalatha et al., 2015).

Paenibacillus terrae isolate ME27-1 exhibited its highest carboxymethylcellulase (CMCase) activity in liquid culture (Liang et al., 2014).

Table 4. Distribution of cellulase producing microorganisms from various sources

Species	Sources/locations	References
Bacteria:		
Anoxybacillus flavithermus EFP1, *Geobacillus thermodenitrificans* EFP2 and *Geobacillus stearothermophilus* EFP3	Water and soil samples, Egyptian hot spring, Hammampharoan, in the Sinai desert in the east of Egypt at the Red Sea shore	Ibrahim and El-Diwany, 2007
Paenibacillus cellulositrophicus	Soil from Pua district in Nan Province, Thailand	Akaracharanyaet al., 2009
Cellulosimicrobium cellulans	Decaying wood	Song and Wei, 2010
Cellulomonas sp. YJ5	Soil from northern Taiwan	Yin et al., 2010
Streptomyces viridobrunneus SCPE-09	Soilunder a sugar cane plantation in Brazil	Da Vinha et al., 2011
Paenibacillus sp. P118	Gut of the tropical fish, Mato Grosso River (Saquarema, Rio de Janeiro, Brazil)	de Castro et al., 2011
Vibrio, *Acinetobacter*, *Providencia*, *Yersinia*, *Pseudomonas*, *Morganella* and *Aeromonas*	Gut of herbivorous grass carp (*Ctenopharyngodon idella*)	Jiang et al., 2011
Shigella flexneri G3	Rumen fluid sample from slaughterhouse in Harbin, China	Wang et al., 2011
Aneurinibacillus thermoaerophilus WBS2	Water from Bakreshwar Hot Ppring, West Bengal, India	Acharya and Chaudhary, 2012a
Klebsiella variicola RBEB3	Kitchen waste soil and garden soil	Gopinath et al., 2012
Cellulomonas sp. ASN2	Soil from Kasur, Pakistan	Irfan et al., 2012; Safdar et al., 2013
Paracoccus pantotrophus FMR19	Goat rumen fluid	Faridha et al., 2013
Mesorhizobium sp.	Marine environment sample at the coastal areas of Cuddalore, Pondicherry, Tamilnadu and Mangalore, Karnataka, India	Prasad and Sethi, 2013
Micrococcus spp., *Bacillus* spp., *Pseudomonas* spp., *Xanthomonas* spp. and *Brucella* spp.	Mangrove soils from Mahanadi River Delta, Odisha, India	Behera et al., 2014
Bacillus tequilensis PH3-48, *B. cereus* PJ1-24B, *B. altitudinis* PH27, *B. safensis* PJ1-24 and *Lysinibacillus fusiformis* PT31	Oil palm meal samples from Phetchaburi, Prachuap Khiri Khan and Pattani Provinces, Thailand	Khianngam et al., 2014
Paenibacillus terrae ME27-1	Soil from Maoer Mountains (Guilin City), China	Liang et al., 2014
Bacillus cereus, *B. subtilius* and *B. thuringiensis*	Soil from the Botanical Garden, Karnatak University Campus, Karnataka, India	Patagundi et al., 2014
Acinetobacter sp. KKU44	Rice straw	Siripornadulsil et al., 2014

Species	Sources/locations	References
*Bacillus subtilis*BY-2	Tibetan pig's intestine	Yang et al., 2014
Bacilluscereus BR0302	Soils from Bueng Samnak Yai (Nong Chamrung) Coastal Wetlands, Rayong Province, Thailand	Chantarasiri et al., 2015
*Bacillus*sp.TS5SRP and *Ochrobactrum anthropic* TS16MCN	Soil near a timber depot in Guntur District, A.P., India	Duza and Mastan, 2015
Bacillus vallismortis RG-07	Soil	Gaur and Tiwari, 2015
*Isoptericola variabilis*IDAH9	Soil from a hot spring of Dehloran in the Ilam Province in the of Southwest of Iran	Azizi et al., 2015
Enhydrobacter sp. ACCA2	Leaf-litter compost from Western Ghats region of Madurai, India	Premalatha et al., 2015
Fungi:		
*Rhizopus oryzae*IMI 298280	Stored potato tubers (*Solanum tuberosum* L.)	Amadioha, 1993
Talaromyces emersonii	Decomposing lignocellulosic materials in Bangladesh	Gomes et al., 2007
Aspergillus niger DPK-cl-12, *Gliomastix murorum* var. *murorum* (= *Acremonium murorum*) DPK-cl-44 and *Stachybotrys chartarum* DPK-cl-111	Waste paper and waste water from the "Klaipėdoskartonas" cardboard factory located in Lithuania	Pečiulytė, 2007
*Penicillium*sp. CR-313 and *Penicillium* sp. CR-316	Soil from the subtropical forest, Puerto Iguazú, Argentina	Picart et al., 2007
Trichoderma asperellum UPM1 and *Aspergillus fumigatus*UPM2	Rotten oil palm empty fruit bunch	Abu Bakar et al., 2010; Ibrahim et al., 2013
Mucor hiemalis f. corticola	Agricultural soil from Iran	Lotfi et al., 2010
Chaetomium sp. NIOCC 36	Wood pieces from mangrove swamps of Sal, Chorao and Zuari in Goa, India	Ravindran et al., 2010
Alternaria sp. MS28	Soil	Sohail et al., 2011
Aspergillus fumigatus M.7.3 and *Myceliophthora thermophilea* M.7.7	A sugar cane bagasse pile in the northwest of São Paulo State, Brazil	Moretti et al., 2012
Fusarium oxysporum H57-1	Flax plant infected by fungal wilt	Lin et al., 2012
*Thermoascus aurantiacus*RBB1	Compost soil	Dave et al., 2013
Sporotrichum thermophile LAR5	Agricultural waste	Bajaj et al., 2014
Aspergillus niger, *A. fumigatus*, *Mucor mucedo* and*Penicillium chrysogenum*	Decayed wood cuttings from a saw mill located at Uwasota Road, Benin City, Edo State, Nigeria	Emoghene et al., 2014
Aspergillus fumigatus, *A. niger* and *A. terreus*	*Artemisia annua* plantation soil in Gangnum, Langtang South local government area of the Plateau State, Nigeria	Ogbonna et al., 2015

3.2. Actinomycetes

Actinomycetes distributed in soils are potential cellulase-producers and are involved in the degradation of organic matter in compost and related materials. *Streptomyces cellulolyticus* LXT, which decomposes cellulose, was proposed as a new cellulolytic species (Li, 1997). *Streptomyces* sp. S7 from decayed fruit waste exhibited its highest crude cellulase enzyme production at pH 5.0 and a temperature of 40°C in a medium that was supplemented with fruit waste as a carbon source (Rathnan and Ambili, 2011) while the strains from marine sediments of the south Indian coastal region exhibited maximal amylase and cellulaseproduction (Selvam et al., 2011). *S. viridobrunneus* SCPE-09 from a soil sample under a sugar cane plantation in Brazil showed that media containing wheat bran 2.0% (w/v) and corn steep liquor 0.19% (w/v) led to their highest cellulase production, 2.0 U/mL of CMCase, when fermented for five days (Da Vinha et al., 2011). *S. griseorubens* St-1 from soil showed an optimal growth at pH 7.0, temperature of 45°C and 6 days of incubation.It also showed maximal levels of reducing sugar production from filter paper hydrolysis at these optimal parameters (Prasad et al., 2013). This strain of actinomycetes showed a protein with a molecular weight very similar to that of *Trichoderma viride* cellulase and *Trichoderma reesei* endoglucanaseas determined by SDS-PAGE (Pirzadah et al., 2014). *S. griseochromogenes* 51, *S. rochei* 157, *Streptomyces* sp. 194 and *S. albidoflavus* NRRL B-16746 showed appreciable zones of clearance indicating cellulase activity (Das et al., 2014). Actinomycetes from sediment samples from the Bhitarkanika mangrove forest exhibited extracellular cellulase activities ranging from 0.266 ± 0.001 to 0.734 ± 0.001 IU/mL for FPC and 0.501 ± 0.014 to 1.381 ± 0.024 IU/mL for CMCase or endoglucanase (Mohanta, 2014). Most of the *Streptomyces*, *Clostridium* and *Bacillus* strains isolated from coffee exocarp samples in coffee-producing areas in Vietnam exhibited cellulose-degrading ability on filter paper (Bui, 2014).

3.3. Fungi

Fungi have been exploited to produce a wide variety of cellulases and hemicellulases. They have the capability to produce large amounts of cellulolytic enzymes which are often less complex than bacterial cellulases. This makes extraction and purification easier. *Aspergillus niger* DPK-cl-12, *Acremonium murorum* DPK-cl-44 and *Stachybotrys chartarum* DPK-cl-111 produced potential exocellular cellulases (overall FPClase and endoglucanase – CMCase). Cellulase activity of these fungal strains determined under study conditions was higher than those of *Trichoderma* strains, and are known as high cellulase producing fungi (Pečiulytė, 2007). *Penicillium* sp. CR-316 and CR-313 from soils secreted a high levels of cellulase in media supplemented with rice straw. Carboxymethylcellulase activity of *Penicillium* sp. CR-316 was optimal at 65°C and pH 4.5. Its activity remained stable after incubation at 60°C and pH 4.5 for 3 h. Crude cellulase produced by *Penicillium* sp. CR-316 showed activity and stability at high temperatures. *Penicillium* sp. CR-316 produced a thermostable cellulase that makes it a good candidate for industrial applications (Picart et al., 2007). The thermophilic fungus, *Talaromyces emersonii,* produced high levels of cellulase in shaken cultures using sugar cane bagasse as a substrate (Gomes et al., 2007). Some genera of Zygomycetes, *Mucor hiemalis* f. *corticola* had high sugar production and *Mucor circinelloides* f. *circinelloides* had the lowest sugar levels (Lotfi et al., 2010). *Aspergillus*, *Thermomyces*, *Myceliophthora*, *Thermomucor* and *Candida* strains isolated from soil, decaying organic matter and sugarcane piles produced xylanase and cellulases in solid state fermentation (SSF). The highest cellulase and xylanase productions were obtained by the cultivation of the strains identified as *Aspergillus fumigatus* M.7.1 and *Myceliophthora thermophila* M.7.7 (Moretti et al., 2012).

Alkaline cellulases were screened from the fifty four fungal isolates obtained from mangrove leaves and mangrove wood litter as endophytes and wood litter fungi, respectively. Only three fungal isolates tested positive for alkaline cellulases under plate assay, among which *Chaetomium* sp. (NIOCC 36) was found to grow over a wide pH range (4.0 to 12.0). This alkaline tolerant fungus was further tested for the production of alkaline cellulases (β-endoglucanase, β-exoglucanase, β-glucosidase), using agricultural and industrial wastes as substrates in submerged (SMF) and solid state fermentation (SSF) processes. Cotton seed under SSF condition showed maximal enzyme production at highly alkaline pH values. The results revealed that cellulase enzymes produced under alkaline conditions using cotton seed in SSF had higher stability and activity than those produced under non-alkaline conditions, i.e., pH 5.0, 7.0, 12.0 and at 50°C. The agricultural and industrial wastes used as carbon sources are inexpensivefor commercial production of cellulase (Ravindran et al., 2010). *Alternaria* sp. MS28 produced different cellulase activies (Sohail et al., 2011).

A thermophilic fungus, *Thermoascus aurantiacus* RBB1, was isolated from composting soil and was optimised for production of cellulolytic enzymes (endoglucanase, filter paper cellulase by SSF) (Dave et al., 2013). *Sporotrichum thermophile* LAR5 successfully produced cellulases by utilizing low-cost agricultural residues such as wheat bran (2000 IU/L), maize bran (1800 IU/L) and rice husks (1600 IU/L) (Bajaj et al., 2014). *Fusarium oxysporum* Schl. H57-1 produce its highest cellulase activity, 1.43 IU/mL. The most suitable pH for CMCase was 4.5-5.5, and the most suitable temperature was 60-65°C. Its activity was stable at pH 5.3-9.5 and 4-50°C (Lin et al., 2012). *P. chrysogenum* isolated from the cuttings of a saw mill located at Uwasota Road, Benin City using serial dilution and pour plate techniques, exhibited its highest cellulolytic potential amongst the fungal isolates, while *Mucor mucedo* had the least cellulolytic activity during the screening (Emoghene et al., 2014). *Aspergillus fumigatus, A. niger, A. terreus* were isolated from *Artemisia annua* L. plantation soil. *A. niger* exhibited its highest cellulase activity on CMC and filter paper while *A. fumigatus* exhibited its optimal activity on filter paper and cellulose acetate media. *A. terreus* had a cellulytic activity on a cellulose acetate medium(Ogbonna et al., 2015).

4. Production of Cellulase by Microorganisms

4.1. Bacteria

In general, cellulases are divided into 3 groups based on how they act on cellulose as a substrate; endoglucanases (EG; EC 3.2.1.4), cellobiohydrolases (CBH; EC 3.2.1.91), and β-glucosidases (G; EC 3.2.1.21) (Sambasivarao et al., 2014). The major source of cellulolytic activity was from bacteria cultivated on a medium containing carboxymethylcellulose sodium salt (CMC-Na) as a substrate for endoglucanase production. Cellulases were induced by a wide variety of substrates, such as low cost agriculture to increase the high yield of enzyme. Furthermore, the sources of carbon, nitrogen and metal ions as well as pH and temperature had effect on the cellulose production. Cellulase productivity of various bacteria is shown in Table 5.

Bacillus strains have been reported to produce cellulose using various sources of carbon and nitrogen, pH and temperature. *Bacillus* sp. C1AC55.07 produced its highest carboxymethylcellulase activity (366 U/L) in 10.0 g/L CMC, 2.0 g/L yeast extract, 30% inoculum (v/v), and temperature of 32°C at 54 h after optimization (Dias et al., 2014). *B. subtilis* isolated from sweet potato *(Ipomoea batatas.* (L.) Lam.) peels was optimized for cellulolytic activities at different pHs (3.0, 4.0, 5.0, 6.0 and 7.0), temperatures (25, 37, 45, 55 and 65°C). After 24 to 48 h of incubation, *B. subtilis* had its best cellulase production of 0.75 U/mL at 55°C, pH 7.0 using CMC as a substrate (Atinuke and Samuel, 2015).

Table 5. Bacerial cellulase production

Species	Enzyme production	Carbon source	pH	Temp. (°C)	References
Clostridium papyrosolvens CFR-703	35 IU/mL	Coconut coir	6.5	35	Swaroopa Rani et al., 2004
Streptomyces drozdowiczii	595 U/L	CMC-Na	7.0	30	de Lima et al., 2005
Anoxybacillus flavithermus EHP1	65.50 U/mL	CMC-Na	-	70	Ibrahim and El-Diwany, 2007
Enterobacter cloacae	0.9 U/mL	CMC-Na	-	37	Sami et al., 2008
Streptomyces sp. J2	432 U/L	CMC-Na	7.0	60	Jaradat et al., 2008
Bacillus pumilus EWBCM1	0.585 IU/mL	CMC-Na	6.0	37	Shankar and Isaiarasu, 2011
Paenibacillus polymyxa	7.814 U/mg	Mango peel	5.5	37	Kumar et al., 2012
Cellulomonas sp. ASN2	0.454 IU/mL	Potato waste	7.0	35	Safdar et al., 2013
Achromobacter xylosoxidans BSS4	68.37 U/mL	CMC-Na	7.0	40	Sreedevi et al., 2013
Bacillus sp. BSS3	104.68 U/mL	CMC-Na	9.0	37	Sreedevi et al., 2013
Pseudomonas sp. BSS2	91.28 U/mL	CMC-Na	8.0	37	Sreedevi et al., 2013
Mesorhizobium sp.	3.67 U/mL	Eucalyptus	9.0	40	Prasad and Sethi, 2013
Bacillus sp. C1AC55.07	366 U/L	CMC-Na	-	32	Dias et al., 2014
Bacillus subtilis BY-2	3.560 U/mL	CMC-Na	5.5	37	Yang et al., 2014
Bacillus subtilis	0.75 U/mL	CMC-Na	7.0	55	Atinuke and Samuel, 2015
Bacillus sp. Y3	7.82 IU/ mL	CMC-Na	7.0	37	Lugani et al., 2015
Micrococcus sp.	0.9490 IU/L	CMC-Na	8.0	37	Nisha, 2015
Enhydrobacter sp. ACCA2	8.86 U/mL	CMC-Na	6.5	30	Premalatha et al., 2015
Sphingomonas sp. FLX-7	16.95 U/mL	CMC-Na	7.2	10.4	Li et al., 2016
Streptomyces viridobrunneus SCPE-09	2.0 U/mL	Wheat bran	7.0	28	Da Vinha et al., 2011

Shankar and Isaiarasu (2011) determined the cellulase production of *B. pumilus* EWBCM1 and found a higher enzyme activity of 0.5851 ± 0.006 IU/mL under submerged fermentation (SMF) after 72 h. The best conditions for *B. pumilus* EWBCM1 was found to be 37°C, pH 6.0 and a medium containing galactose, malt extract, ammonium molybdate, calcium chloride, 2.5% NaCl, and Tween-20 using a 2% inoculum. Acharya and Chaudhary, (2012b) studied the effect of some nutritional and environmental factors on the production of endoglucanase (CMCase) and exoglucanases (FPase) from *B. licheniformis*MVS1 and *Bacillus* sp. MVS3. Both isolates showed their maximal cellulase production after 60 h of incubation using wheat and rice straw as substrates at pH 6.5-7.0 incubated at 50-55°C. Furthermore, FPase production was much higher than CMCase production by both of the strains. *B. subtilis* BY-2 produced its maximal cellulase (3.56 U/mL) with a medium containing with 1% corn flour as the carbon source and 2% peptone and yeast powder as the nitrogen source. The initial pH during fermentation was 5.5 at 37°C on a rotary shaker at 220 rpm, with a 4% inoculum, resulting in a high and stable amount of enzyme in 24-48 h (Yang et al., 2014).

Lugani et al., (2015) reported that the maximal FPase and CMCase activities of *Bacillus* sp. Y3 were 6.84 IU/mL and 7.82 IU/mL, respectively, when a basal media at pH 7.0 containing CMC 1% (w/v) and peptone 1% (w/v) was inoculated with a 2% (v/v) inoculum and incubated at 37°C for 96 h. *Clostridium papyrosolvens* CFR-703 produced a high yield of carboxymethyl cellulase under anaerobic cultivation at 35°C (35 IU/mL) using 0.5% coconut coir as a substrate after 48 h (Swaroopa Rani et al., 2004). Furthermore, yeast extract, Fe^{2+} and at pH 6.5 had an effect to support carboxymethyl cellulase production. Cellobiase activities were optimal at 35°C, pH 7.5, when absorbent cotton, yeast extract and Mg^{2+}were used in the culture medium. *Anoxybacillus flavithermus* EHP1 showed a higher of rate of CMC degradation (enzyme production, 65.50 U/mL) at 70°C, 180 rpm at the end of the stationary phase (36 h) than on avicel, cellobiose and xylan (Ibrahim and El-Diwany, 2007).

Sami et al., (2008) studied the effect of different carbon sources including carboxymethyl cellulose (CMC) and 2% avicel on the production of endo-1, 4-β–D-glucanases by*Enterobacter cloacae*. This bacterium produced maximal levels of cellulase(0.9 U/mL) after 96 h of fermentation using CMC as substrate while incubated on orbital shaker at 100 rpm at 37°C.*Paenibacillus polymyxa* produced its maximal CMCase, 7.814 U/mg, in a medium containing 7% mango peel (w/v) and 1.5% ammonium sulphate (w/v) at 37°C and pH 5.5 (Kumar et al., 2012) compared to *Cellulomonas* sp. ASN2. This microorganism was isolated from soil and showed its maximalcellulase production (0.454 IU/mL/min) with initial medium pH of 7.0, inoculum size of 2% (v/v) and incubation temperature of 35°C for 48 h (Safdar et al., 2013). The optimal medium for producing enzyme was supplemented with 2% corn steep liquor, using potato waste as a substrate in a submerged fermentation.

Sreedevi et al., (2013) observed the influence of growth parameters such as pH, temperature, CMC concentration and agitation on cellulase production from *Achromobacter xylosoxidans* BSS4, *Bacillus* sp. BSS3 and *Pseudomonas* sp. BSS2. *Achromobacter xylosoxidans* BSS4 exhibited a maximalcellulaseproduction of 68.37 U/mL, at pH 7.0, 40°C with 0.5% CMC and 150 rpm. The highest enzyme production from *Bacillus* sp. BSS3 was

104.68 U/mL, at pH 9.0, 37°C with 1% CMC, 150 rpm.*Pseudomonas* spshowed optimal cellulase production at pH 8.0, 37°C with 1% CMC and 150 rpm. *Pseudomonas* sp. BSS2 increased its enzyme production from an initial concentration of 49.84 U/mL to 91.28 U/mL after optimization. *Mesorhizobium*sp. isolated from a marine source produced it highestcellulaselevel in the pH range of 5.0-9.0 using eucalyptus as a substrate, ammonium nitrate as a nitrogen source at 40°C. Cellulase activity was lowest when paper was used as a substrate. Additionally, maximalenzymatic activity was observed when the substrate was treated at 200°C for 1 h with 0.9-1% acid or alkali (Prasad and Sethi, 2013).

Nisha (2015) reported that *Micrococcus* sp. was optimized using environmental factors such as incubation temperature, pH, various concentrations of CMC and incubation times. Maximalcellulase activity was found to be 0.9490 IU/L using 1.5% CMC as substrate, incubating at 37°C for 72 h at pH 8.0. Premalatha et al., (2015) studiedcellulase production of*Enhydrobacter* sp. ACCA2. Optimized medium composition was 1.5% carboxymethylcellulose, 1.445% peptone, 0.044% NaCl, 0.031% $MgSO_4$, 0.15% K_2HPO_4, and 0.045% $(NH_4)_2SO_4$, which yielded up to a 2.39 fold increase in cellulase activity at pH 6.5, 30°C, with 3 days of incubation. *Sphingomonas* sp. FLX-7, a psychrotrophic cellulosedegrading bacterium produced maximalcellulase activity, 16.95 U/mL, using a 2.1% inoculum, 73.5 h of incubation at pH 7.2. This experiment used CMC as a substrate for cellulase production and the culture was incubated at 10.4°C (Li et al., 2016).

Streptomyces drozdowiczii was cultivated at 30°C using either carboxymethylcellulose (CMC), wheat bran (WB) or distillers dried grains (DDG) as carbon sources, and either yeast extract (YE) or corn steep liquor (CSL) as nitrogen sources. The highest CMCase level of 595 U/L was produced from a medium containing 0.3% (w/v) YE and 1% (w/v) CMC when the culture was incubated at 30°C, at pH 7.0, under agitation (200 rpm), for 5 days (de Lima et al., 2005). Similarly, *Streptomyces* sp. J2 was grown in a CMC broth and showedits maximal enzyme activity (432 U/L) in a medium that was supplemented with 0.5% glucose, 0.2% starch, and 0.2% NH_4Cl at pH 7.0 after 3 days of incubation. However, 60°C was found to be the optimal temperature (Jaradat et al., 2008). Furthermore, *S. viridobrunneus* SCPE-09 was tested with different kinds and concentrations of carbon (sugar cane bagasse, wheat bran) and nitrogen source (steep liquor). The results showed that media containing 2.0% (w/v)wheat bran and corn steep liquor at0.19% (w/v) led to the highest yield, 2.0 U/mL of CMCase, obtained on the fifth day of fermentation. The culture medium was adjusted to pH 7.0, incubated at 28°C and shaken (200 rpm) (Da Vinha et al., 2011).

4.2. Fungi

Cellulase fermentations by fungi were performed in solid state fementation (SSF) or submerge fermentation (SmF) depending on the optimal condition of each of strain. Commonly, fungi produce high level of cellulase using various substrates and medium components (Table 6). Moosavi-Nasab and Majdi-Nasab (2007) studied cellulase production by *Trichoderma reesei*, using sugar beet pulp (SBP) as a substrate. SBP is a by-product of the beet sugar industry that contains large amounts of cellulose. The maximum filter paper activity was 0.46 IU/mL using a medium containing a mixture of basal salts and SBP substrate with a fermentation time of 4-6 days, at 30°C. A *T. viride* strain produced maximal cellulase activity when it used 4-5% (w/v) of municipal solid waste residue and 1.0% (w/v) of yeast extract as its carbon and nitrogen sources, respectively. Municipal solidwaste residue was best for exoglucanase (1.77 U/mL), endoglucanase (1.95 U/mL) and β–glucosidase (1.66 U/mL) production. Optimum pH of the medium, temperature and incubation period for production of the cellulolytic enzyme was 6.5, 45°C and for 6 days, respectively (Gautam et al., 2010).

Aspergillus strains have been reported to produce cellulase using agricultural and industrial wastes as substrates. *Aspergillus niger* used sawdust as a substrate for cellulase production (Acharya et al., 2008). Alkaline pretreated (2 N NaOH) sawdust at a 9.6% concentration gave 0.1813 IU/mL of cellulase activity. The highest enzyme activities were obtained under submerged fermentation at 120 rpm, at 28°C for 96 h. The medium for best cellulase activity contained 0.125% peptone, 0.14% ammonium sulphate, 0.03% urea and had a pH range of 4.0-4.5. Mrudula and Murugammal (2011) studied cellulase production in submerged (SmF) and solid state (SSF) fermentation by *Aspergillus niger*. Their results showed that cellulase activity in SSF was 14.6 fold higher than in SmF. The CMCase and FPase activities obtained in SSF were 8.89 and 3.56 U/g of dry mycelial bran (DBM) after 72 h, at 30°C and at a moisture ratio of 1:2 (w/v). CMCase and FPase activities from SmF were 3.29 and 2.3 U/mL in the culture broth after 96 h at 30°C. This high yield of CMCase and FPase in both fermentations was found when using coir waste as a substrate, lactose as a carbon source, peptone as a nitrogen source at pH 6.0-7.5.

Table 6. Fungal cellulase production

Species	Enzyme production	Carbon source	pH	Temp. (°C)	References
Trichoderma reesei	FPase (0.46 IU/mL) in SmF	Sugar beet pulp	4.8	30	Moosavi-Nasab and Majdi-Nasab, 2007
Aspergillus niger	CMCase (0.1813 IU/mL) in SmF	Sawdust	4.0-4.5	28	Acharya et al., 2008
Trichoderma viride	Exoglucanase (1.77 U/mL), endoglucanase (1.95 U/mL) and β–glucosidase (1.66 U/mL) in SmF	Municipal solid wasteresidue	6.5	45	Gautam et al., 2010
Aspergillus niger	CMCase (8.89 U/g) and FPase (3.56 U/g) in SSF, CMCase (3.29 U/g) and FPase (2.3 U/g) in SmF	Coir waste	6.0-7.5	30	Mrudula and Murugammal, 2011
Alternaria sp. MS28	Endoglucanase (0.75 IU/mL) and β-glucosidase (0.05 IU/mL) in SmF	CMC-Na and phosphoric acid swollencellulose (PASC) orfilter paper	-	35	Sohail et al., 2011
Aspergillus terreus	endoglucanase (213. 3 IU/mL) in SmF	Lantana leaves	3.0	37	Padmavathi et al., 2012
Mucor plumbeus	endoglucanase (206.0 IU/mL) in SmF	Lantana leaves	3.0	37	Padmavathi et al., 2012
Fusarium oxysporum JRE1	CMCase (16.8 U/mL) in SmF	CMC-Na	5.6	28	Dar et al., 2013
Schizophyllum commune KUC9397	β-Glucosidase (43.51 U/mL) in SmF	Cellulose	-	27	Lee et al., 2014
Aspergillus terreus MS105	Endoglucanase (3.3604 IU/L/h) and β-glucosidase (8.301 IU/L/h) in SmF	Grass	5.5	35	Sohail et al., 2016

Sohail et al., (2016) screened *Aspergillus terreus* MS105 for cellulase production using various carbon sources, including carboxymethyl cellulose (CMC), avicel, sigmacell, filterpaper and salicin. Cellulase production under SmF of grass was 1.7-20 fold higher than sugarcane-bagasse, corncob and commercially available purified substrates. The maximum endoglucanase and β-glucosidase production from *A. terreus* MS105 was 3.3604 and 8.301 IU/L/h, respectively,in a shaking incubator at 35°C, 200 rpm. Padmavathi et al., (2012) determined production of cellulase using SmF by marine fungi (*A. terreus* and *Mucor plumbeus*). Rice bran, wheat bran, bamboo leaves, banana leaves, peepal leaves, sugar cane leaves, lantana leaves, ragi straw, maize leaves and eucalyptus leaves were used as substrates for the cellulase production. The best enzyme activity of *A. terreus* and *M. plumbeus* was213.3 IU/mL and 206 IU/mL, respectively, which were obtained when Lantana leaves were used as a carbon source. The medium contained Lantana leaves, ammonium sulphate as nitrogen source, at pH 3.0 and was incubated at 37°C for 7 days and showed good cellulase activity with *A. terreus* and *M. plumbeus*.

Alternaria sp. MS28 produced endoglucanase and β-glucosidase when induced with 1% CMC and 1% salicin, respectively. The levels of endoglucanase were, in general, 4-8 times higher than β-glucosidase. Maximal endoglucanase of 0.75 IU/mL was obtained on a mineral salt medium (MSM) and 1% CMC at 35°C. Phosphoric acid swollen cellulose (1% PASC) or 1% filter paper was added to MSM and showed 0.05 IU/mL of β-glucosidase activity (Sohail et al., 2011). *Fusarium oxysporum*, an endophytic fungus, produced its best CMCase activity with 2% CMC (16.8 U/mL) under agitation at 28°C and at pH 5.6 (Dar et al., 2013). Lee et al., (2014) showed β-glucosidase production by *Schizophyllum commune* KUC9397. β-Glucosidase activity in the optimized medium was increased 7.2 fold compared to activity in an optimized medium. The highest β-glucosidase production (43.51 U/mL) was found with 2.96% cellulose, 2.30% soy peptone and 0.11% thiamine HCl with incubation at 27°C, 150 rpm for 4 days.

5. Benefits of Cellulases

Microbial cellulases have become the focal biocatalysts due to their complex nature and wide spread industrial applications. Cellulases are inducible enzymes synthesized by a diversity fungi and bacteria during their growth on cellulosic materials. These enzymes have numerous industrial application including biofuel production, pulp and paper, textiles, laundry, brewing, food and feed industries and agriculture (Kuhad et al., 2011; Menendez et al., 2015).

5.1. Biofuel Production

The degradation of lignocellulosic materials for biofuel production consists of 3 steps: 1) physicochemical pretreatments, 2) enzymatic hydrolysis of cellulose, and 3) fermentation. Cellulases have a potential role in the bioconversion of renewable lignocellulosic biomass involving the hydrolysis of plant biomass. Nowadays, reduction in costs is required to ensure the permanence and viability of commercial cellulase production. Producing inexpensive sugars for biofuel production depends upon sourcing inexpensive feedstocks and using efficient pretreatment and enzymatic hydrolysis steps (Xin et al., 2013; Menendez et al., 2015). Rathna et al., (2014) presented bioethanol production using fungi and yeast. High yield of ethanol was obtained from sawdust by saccharification and fermentation using*Aspergillus ochraceus* and *Saccharomyces cerevisiae*, respectively. *A. niger* (11.5 g/L of ethanol) and *Mucor racemosusresenius* (7.2 g/L of ethanol) were evaluated to produce cellulase using the agro-industrial residue, sugarcane bagasse as (Fischer et al., 2014). Furthermore, Belal (2013) described rice straw as a goodstarting material for bioethanol production by *Trichoderma reesei*. Their ethanol yield was about 10 to 11 g/L. Ezebuiro et al., (2015) studied bioethanol production by *B. cereus* GBPS9 using sugarcane bagasse (18.40 g/L) and cassava peels (17.80 g/L) as feedstocks. Additionally, cassava stalk is an agricultural by-product that can be used as a carbon source for bioethanol production. Saccharification of this lignocellulosic biomass and its fermentation to bioethanol were performed by *B. methylotrophicus* RYC01101 and *S. cerevisiae* TISTR 5111, respectively (Chantarasiri, 2014).

5.2. Pulp and Paper, and Textile Industry

Cellulases are useful for de-inking of different types of paper wastes. They were applied for the release of ink from fiber surfaces by partial hydrolysis of carbohydrate molecules. Moreover, cellulases have potential roles in improving fiber brightness, strength properties and pulp freeness (Kuhad et al., 2011). Ko et al., (2010) described the freeness of fully bleached eucalypt kraft pulp as a function of Cel-BL11 cellulase pretreatment. The cel-BL11 gene from *Paenibacillus campinasensis* BL11 was cloned and expressed in *Escherichia coli*. Das et al., (2013) reported cellulolytic enzyme production by *Aspergillus fumigatus* ABK9 on a wheat bran-rice straw mixed substrate and use of cocktail enzymes for deinking waste office paper pulp. This crude enzyme was able to de-ink waste laser printed paper within 6 h.

Generally, cellulases are used in the textile industry for the modification of cellulosic fibre properties with the aim of improving the quality of fabrics such as softening, defibrillation, and processes for providing localized variation in the colour density of fibers. Many reports showed the application of cellulases to replace the use of pumice stones for biostoning and release of excess indigo dye of denim garments to provide softness and a faded look, and to digest fiber ends protruding from the fabric for a better finish (Shweta, 2012). Acid cellulase (*Trichoderma reesei*) and neutral cellulase (*Humicola isolens*) were used to wash of denim, which has the drawback of back staining the indigo dye on to the fabric. Anish et al., (2007) tested an alkali stable endoglucanase from alkalothermophilic *Thermomonospora* sp. (T-EG) for denim biofinishing under alkaline conditions. The results showed that this strain is effective in removing hairiness with negligible weight loss, imparting softness to the fabric.

5.3. Laundry and Detergents Industry

Cellulases are used in household laundry detergents for improving cleanliness, softness and brightness of fabrics made of cotton, including antiredeposition of ink particles (Bhat, 2000; Shweta, 2012). Furthermore, the important performance characteristics ofcellulases for detergent applications are compatibility with other ingredients present in formulations, temperature stability, and activity under alkaline conditions (Menendez et al., 2015). Ladeira et al., (2015) described the production of cellulases by the thermophilic *Bacillus* sp. SMIA-2 using sugarcane bagasse and corn steep liquor as substrates for industrial applications. An avicelase and CMCase producing strain was studied for properties such as optimum enzyme activity,pH stability and temperature stabilty. In culture medium, this strain released protease, but the cellulase producing strain was not sensitive to protease digestion. Moreover, cellulases were tested to determine their compatibility with each laundry detergent tested. The properties obtained by

Bacillus sp. SMIA-2 suggest that this strain might become a potential source of lignocellulose-degrading enzymes for the detergent industry.

5.4. Brewery, Food Production and Feed Industry

In fermentation, cellulases play principal roles in alcoholic beverages (beers and wines). β-Glucosidase in wine production can enhance must clarification and improve the aroma of wine by modifying glycosylated precursors. Benefits of enzymatic activity during wine making are better maceration, improved color extraction, easy clarification, easy filtration, improved wine quality, and improved stability. For beer brewing, cellulases improve the primary fermentation and quality of beer (Bhat, 2000; Kuhad et al., 2011). β-Glucosidase from *Aspergillus* supplemented with*Trichoderma reesei*cellulase preparations, played an important role for cellulose saccharification on an industrial scale (Kaur et al., 2007). Sources of β-Glucosidase were from yeast (*S. cerevisiae* and *Pichia etchellsii*), mesophilic fungi (*Trichoderma harzianum* and *Aspergillus*sp.) and thermophilic fungi (*Thermoascus aurantiacus*, *Chaetomium thermophile*, *Humicola insolens*, *Sporotrichum thermophile and Melanocarpus*sp.).

Cellulases are widely used in food processing. Production of fruit and vegetable juices requires improved methods for extraction, clarification, and stabilization. They were used in part for their macerating enzymes complexes (cellulases, xylanases, and pectinases) to increase the yield of juices. In fruit juice production, enzymatic maceration is used to develop cloud stability, texture and decrease viscosity of the nectars and purees of tropical fruits. Infusion of enzymes can improve texture, flavor, and aroma properties of fruits and vegetables by reducing excessive bitterness. Moreover, enzymes be can used to improve extraction of olive oil and can also release of the antioxidants from fruit and vegetable pomace (Kuhad et al., 2011). Commercial enzymes are employed in the fruit processing industry for clarification of juices, however these enzymes are quite expensive. Hence, selection of cheaper substrates is needed. *B. pumilis* as a marine bacterium was studied for its cellulase production. This cellulase producing strain was tested for sachaarification of banana-agro waste and clarification of apple juice. Clear juice was obtained indicating the effect of cellulase enzymes on juice clarification (Kanmani et al., 2011). Mango peel residue 1% (w/v) as substrate for a cellulase producing *A. niger* was screened. Cellulases obtained from mango peel residue were found to have good potential for juice clarification (Kumar et al., 2015).

In animal feed processing, cellulases are generally needed to enhance the nutritive quality of forages, to improve the feed conversion ratio, and/or to increase the digestibility of a cereal-based feeds (Menendez et al., 2015). Ali et al., (2014) evaluated the fermentation of a local agro-industry waste to produce Cotton Seed Crude Fiber (CSCF) as an alternative nutritious and palatable feed stuff for rabbits. The results showed that cellulase from *Cellulomonas* could improve CSCF quality. Wongputtisin et al., (2014) improved soya bean hull (SBH) qualities using microbial fermentation. *B.subtilis* TK8 isolated from Tua-nao was used for fermented soya bean hull (FSBH) production. After fermentation, the nutritional quality of SBH was improved by an increase in soluble sugars, soluble proteins, crude protein and crude lipid, and a decrease in the content in the raffinose family of oligosaccharides. Hence, SBH fermentation was shownas a potential method of producinga feed ingredient. Lohlum et al., (2014) used corncobs as a waste product in agricultural processing for fermentation by *A. niger*. The results showed an increase in protein, lipid content and mineral content, but crude fiber decreased after fermentation. Therefore, corncob fermentation using *A. niger* is an effective means of enhancing the nutritional quality of corncobs for use as an energy source in livestock feed formulations.

5.5. Plant Biocontrol

Cellulases have potential roles in agriculture for controlling plant pathogens and diseases. Bacterially secreted cellulases have the ability to lyse cell walls of pathogenic fungi. Cellulolytic fungi and bacteria were applied in agriculture by facilitating enhanced seed germination, rapid plant growth and flowering, improved root system, increased crop yields and enhanced soil quality and reduced dependence on mineral fertilizers (Tejada et al., 2008; Kuhad et al., 2011; Shweta, 2012). *Streptomyces rubrolavendulae* (Yen) S4 was used to control the seedling damping-off disease of the horticultural plant, Joseph's Coat, caused by *Pythium aphanidermatum* (Edson) Fitz. The modes of the antagonistic action of*P. aphanidermatum*involved itshigh production ofcellulase (Loliam et al., 2013). *B. subtilis* displayed high antagonistic activity against *Colletotrichum gloeosporioides* OGC1. Microscopic observations showed a clear hyphal lysis and degradation of fungal cell walls as a result of chitinase, glucanase and cellulose activities (Ashwini and

Srividya, 2014). *Serratia* sp. (MSK1 and MSK24) and *Pseudomonas* sp. (MSK13) are key players for sustaining the soil quality degraded cellulose and exhibiting cellulase activity (Singh et al., 2010).

5.6. Health Benefits

Cellulases have a role in pharmaceutical and medical applications. Since, cellulases like the digestive enzyme products that were used in humans poorly digest cellulose fiber (Karmakar and Ray, 2011). Dietary fibres, including cellulose, are a major source of substrates for the resident anaerobic microbial communities of the colon. Fermentation by gut microbes provides health benefits to human byregulation of the digestive tract, prevention of obesity and diabetes through control of glucose absorption and insulin response, and reduction of cardiovascular diseases and cancer through plant polymer-associated bioactive compounds with anti-oxidant activities (Robert and Bernalier-Donadille, 2003). Cellulolytic bacteria were found in the human colon including *Bifidobacterium*, *Clostridium*, *Bacteroides*, *Faecalibacterium*, *Coprococcus* and *Eubacterium* using PCR and denaturing gradient gel electrophoresis (DGGE as reported by Kopečny et al., (2004).

Cellulases were applied to disrupt pathogenic organisms through mechanisms such as the degradation cell walls of *Acanthamoeba*. This microorgamism can cause blinding keratitis and a fatal brain infection. The cyst wall of this amoeba consists of cellulose, socellulases have a potential therapeutic role in the effective targeting of this trophozoite encased within the cyst walls (Lakhundi et al., 2015). Additionally, the use of cellulose can potentially inhibit *Pseudomonas aeruginosa* biofilms, a pathogenic bacterium commonly found in medical implants (Loiselle and Anderson, 2003). This biofilm consist of microcolonies surrounded by an exopolysaccharide produced by the bacteria.They protect the bacteria against antibiotics and the immune response (Høiby et al., 2001).*Burkholderia cepacia* is an opportunistic pathogen causing infections in patients with cystic fibrosis.It has the ability to grow as a biofilm. Furthermore, a biofilm of *B. cepacia*is involved in contaimination of pharmaceuticals and cosmetics. The potential of the antibiofilm potential of cellulase against *B. cepacia* biofilms to cleave the β-1,4 linkages in exopolysaccharides, a major constituent of biofilms,was found (Rajasekharan and Ramesh, 2013).

In conclusion, cellulases are being commercially produced by several industries and commonly used in many industrial applications. The demand for more stable, highly active and specific enzymes is growing rapidly. Bacteria and fungi have enormous commercial potential in biotechnology and research. Rapid and reliable methods for isolation and screening of cellulase producers from various sources will allow a greater number of novel microbial cellulases for industrial use. These developments together with improved scientific knowledge on the production, purification, characterization, biochemistry and molecular biology of cellulases are expected to pave the way for remarkable success in the biotechnology of cellulases.

REFERENCES

Abu Baker, N. K., Abd-Aziz, S., Hassan, M. A. & Ghazali, F. M. (2010). Isolation and selection of appropriate cellulolytic mixed microbial cultures for cellulases production from oil palm empty fruit bunch. *Biotechnol.*, *9*, 73-78.

Acharya, A., Joshi, D. R., Shrestha, K. & Bhatta, D. R. (2012). Isolation and screening of thermophilic cellulolytic bacteria from compost piles. *Sci. World.*, *10*, 43-46.

Acharya, P. B., Acharya, D. K. & Modi, H. A. (2008). Optimization for cellulase production by *Aspergillus niger* using saw dust as substrate. *Afr. J. Biotechnol.*, *7*, 4147-4152.

Acharya, S. & Chaudhary, A. (2012a). Alkaline cellulase produced by a newly isolated thermophilic *Aneurinibacillus thermoaerophilus* WBS2 from hot spring, India. *Afr. J. Microbiol. Res.*, *6*, 5453-5458.

Acharya, S. & Chaudhary, A. (2012b). Optimization of fermentation conditions for cellulases production by *Bacillus licheniformis* MVS1 and *Bacillus* sp. MVS3 isolated from Indian hot spring. *Braz. Arch. Biol. Technol.*, *55*, 497-503.

Akaracharanya, A., Lorliam, W., Tanasupawat, S., Lee, K. C. & Lee, J. S. (2009). *Paenibacillus cellulositrophicus* sp. nov., a cellulolytic bacterium from Thai soil. *Int. J. Syst. Evol. Microbiol.*,*59*, 2680-2684.

Ali, U. Chuzaemi, S., Soebarinoto & Sjofjan, O. (2014). The effect fermentation of local agroindustry waste using cellulolytic bacteria *Cellulomonas* on nutrient content as feed stuff. *J. Biol. Agr. Healthcare.*,*4*, 68-72.

Amadioha, A. C. (1993). Production of cellulolytic enzymes by *Rhizopus oryzae* in culture and *Rhizopus*-infected tissues of potato tubers. *Mycologia.*, *85*, 574-578.

Anish, R., Rahman, M. S. & Rao, M. (2007). Application of cellulases from an alkothermophilic *Thermomonospora* sp. in biopolishing of denims. *Biotechnol. Bioeng.*, *96*, 48-56.

Apun, K., Jong, B. C. & Salleh, M. A. (2000). Screening and isolation of a cellulolytic and amylolytic *Bacillus* from sago pith waste. *J. Gen. Appl. Microbiol.*, *46*, 263-267.

Ashwini, N. & Srividya, S. (2014). Potentiality of *Bacillus subtilis* as biocontrol agent for management of anthracnose disease of chilli caused by *Colletotrichum gloeosporioides* OGC1. *3Biotech.*, *4*, 127-136.

Atinuke, O. A. & Samuel, F. (2015). Production of bacterial amylases and cellulases using sweet potato (*Ipomoea batatas*. (L.) Lam.) peels. *Afr. J. Biochem. Res.*, *9*, 104-109.

Azizi, M., Hemmat, J., Seifati, S. M., Torktaz, I. & Karimi, S. (2015). Characterization of a thermostable endoglucanase produced by *Isoptericola variabilis* sp. IDAH9. *Braz. J. Microbiol.*, *46*, 1225-1234.

Bajaj, B. K., Sharma, M. & Rao, R. S. (2014). Agricultural residues for production of cellulase from *Sporotrichum thermophile* LAR5 and its application for saccharification of rice straw. *J. Mater. Environ. Sci.*, *5*, 1454-1460.

Banakar, S. P., Thippeswamy, B., Naveenkumar, K. J. & Thirumalesh, B. V. (2012). Isolation and partial purification of cellulases from forest soils of Bhadra wildlife Sanctuary, Western Ghats of southern, India. *Int. J. Curr. Res.*, *4*, 076-082.

Behera, B. C., Parida, S., Dutta, S. K. & Thatoi, H. N. (2014). Isolation and identification of cellulose degrading bacteria from mangrove soil of Mahanadi river delta and their cellulase production ability. *Am. J. Microbiol. Res.*, *2*, 41-46.

Belal, E. B. (2013). Bioethanol production from rice straw residues. *Braz. J. Microbiol.*, *44*, 225-234.

Bhat, M. K. (2000). Cellulases and related enzymes in biotechnology. *Biotechnol. Adv.*, *18*, 355-383.

Bradner, J. R., Gillings, M. & Nevalainen, K. M. H. (1999). Qualitative assessment of hydrolytic activities in Antarctic microfungi grown at different temperatures on solid media. *World. J. Microbiol. Biotechnol.*, *15*, 131-132.

Bui, H. B. (2014). Isolation of cellulolytic bacteria, including actinomycetes, from coffee exocarps in coffee-producing areas in Vietnam. *Int. J. Recycl. Org. Waste. Agricult.*, *3*, 48.

Chantarasiri, A. (2014). Novel halotolerant cellulolytic *Bacillus methylotrophicus* RYC01101 isolated from ruminant feces in Thailand and its application for bioethanol production. *KMUTNB. Int. J. Appl. Sci. Technol.*, *7*, 63-68.

Chantarasiri, A., Boontanom, P., Yensaysuk, N. & Ajwichai, P. (2015). Isolation and identification of a cellulase-producing *Bacillus* sp. strain BR0302 from Thai coastal wetland soil. *KMUTNB. Int. J. Appl. Sci. Technol.*, *8*, 197-203.

Couri, S. & Farias, A. X. (1995). Genetic manipulation of *Aspergillus niger* for increased synthesis of pectinolytic enzymes. *Rev. Microbiol.*, *26*, 314-317.

Crawford, J. H. (1983). Composting of agricultural wastes a review. *Process. Biochem.*, *18*, 14-18.

Da Vinha, F. N., Gravina-Oliveira, M. P., Franco, M. N., Macrae, A., da Silva Bon, E. P., Nascimento, R. P. & Coelho, R. R. (2011). Cellulase production by *Streptomycesviridobrunneus* SCPE-09 using lignocellulosic biomass as inducer substrate. *Appl. Biochem. Biotechnol.*, *164*, 256-267.

Damaso, M. C. T., da Costa Terzi, S., Farias, A. X., de Oliveira, A. C. P., Fraga, M. E. & Couri, S. (2012). Selection of cellulolytic fungi isolated from diverse substrates. *Braz. Arch. Biol. Technol.*, *55*, 513-520.

Damaso, M. C. T., Passianoto, M. A., Freitas, S. C., Freire, D. M. G., Lago, R. C. A. & Couri, S. (2008). Utilization of agroindustrial residues for lipase production by solid state fermentation. *Braz. J. Microbiol.*, *39*, 1-6.

Dar, R. A., Saba, I., Shahnawaz, M., Sangale, M. K., Ade, A. B., Rather, S. A. & Qazi, P. H. (2013). Isolation, purification and characterization of carboxymethyl cellulase (CMCase) from endophytic *Fusarium oxysporum* producing podophyllotoxin. *Adv. Enz. Res.*,*1*, 91-96.

Das, A., Paul, T., Halder, S. K., Jana, A., Maity, C., Das Mohapatra, P. K., Pati, B. R. & Mondai, K. C. (2013). Production of cellulolytic enzymes by *Aspergillus fumigatus*ABK9 in wheat bran-rice straw mixed substrate and use of cocktail enzymes for deinking of waste office paper pulp. *Bioresource. Technol.*, *128*, 290-296.

Das, P., Solanki, R. & Khanna, M. (2014). Isolation and screening of cellulolytic actinomycetes from diverse habitats. *Int. J. Adv. Biotechnol. Res.*, *5*, 438-451.

Dave, B. R., Sudhir, A. P., Parmar, P., Pathak, S., Raykundaliya, D. P. & Subramanian, R. B. (2013). Enhancement of cellulase activity by a new strain of *Thermoascus aurantiacus*: Optimisation by statistical design response surface methodology. *Biocatal. Agr. Biotechnol.*, *2*, 108.115.

de Castro, A. L. M., Vollú, R. E., Peixoto, R. S., Grigorevski-Lima, A. L., Coelho, R. R. R., Bon, E. P. S., Rosado, A. S. & Seldin, L. (2011). Cellulolytic potential of a novel strain of *Paenibacillus*sp. isolated from the armored catfish *Parotocinclusmaculicauda* gut. *Braz. J. Microbiol.*, *42*, 1608-1615.

de Lima, A. L. G., do Nascimento, R. P., da Silva Bon, E. P. & Coelho, R. R. R. (2005). *Streptomyces drozdowiczii*cellulase production using agro-industrial by-products and its potential use in the detergent and textile industries. *Enz. Microbial. Technol.*, *37*, 272-277.

Dias, P. V. S., Ramos, K. O., Padilha, I. Q. M., Araújo, D. A. M., Santos, S. F. M. & Silva, F. L. H. (2014). Optimization of cellulase production by *Bacillus*sp. isolated from sugarcane cultivated soil. *Chem. Eng. T.*, *38*, 277-282.

Dubey, R. C. & Maheshwari, D. K. (2004). Practical microbiology. S. Chand & Company Ltd., New Delhi. 352 pp.

Duza, M. B. & Mastan, S. A. (2015). Optimization studies on cellulase production from *Bacillusanthracis* and *Ochrobactrum anthropic* (YZ1) isolated from soil. *Int. J. Appl. Sci. Biotechnol.*, *3*, 272-284.

El-Nakeeb, M. A. & Lechevalier, H. A. (1963). Selective isolation of aerobic actinomycetes. *Appl. Microbiol.*, *11*, 75-77.

El-Sersy, N. A., Abd-Elnaby, H., Abou-Elela, G. M., Ibrahim, H. A. H. & Toukhy, N. M. (2010). Optimization, economization and characterization of cellulase produced by marine *Streptomyces ruber*. *Afr. J Biotechnol.*, *9*, 6355-63644.

Emoghene, A. O., Okungbowa, O. C., Obayagbona, O. N. & Jaboro, A. G. (2014). Cellulolytic activities of wild type fungi isolated from decayed wood cuttings. *Nig. J. Biotechnol.*, *27*, 41-48.

Ezebuiro, V., Ogugbue, C. J., Oruwari, B. & Ire, F. S. (2015). Bioethanol production by an ethanol-tolerant *Bacilluscereus* strain GBP59 using sugarcane bagasse and cassava peels as feed stock. *J. Biotechnol. Biomater.*, *5*, 4. (doi:10.4172/2155-952X.1000213).

Faridha, B. I., Meignanalaksmi, S. & Pandima, D. M. (2013). Isolation and characterization of cellulase producing *Paracoccus pantotrophus* FMR19 (JX012237) from goat ruman fluid and its effect on pH, temperature and carbon sources. *Int. J. Adv. Biotechnol.*, *4*, 384-390.

Fischer, J., dos Santos Lopes, V., Santos, E. F. Q., Filho, U. C. & Cardoso, V. L. (2014). Second generation ethanol production using crude enzyme complex produced by fungi collected in Brazilian Cerrado (Brazilian Savanna). *Chem. Eng. T.*, *38*, 487-492.

Fortnagel, P., Harms, H., Wittich, R. M., Krohn, S., Meyer, H., Sinnwell, V., Wilkes, H. & Francke, W. (1990). Metabolism of dibenzofuran by *Pseudomonas*sp. strain HH69 and the mixed culture HH27. *Appl. Environ. Microbiol.*, *56*, 1148-1156.

Gaur, R. & Tiwari, S. (2015). Isolation, production, purification and characterization of an organic-solvent-thermostable alkalophilic cellulase from *Bacillus vallismortis* RG-07. *BMC. Biotechnol.*, *15*, 1-12.

Gause, G. F., Preobrazhenskaya, T. P., Sveshnikova, G. V., Terekhova, L. P. & Maksimova, T. S. (1983). *A Guide for Determination of Actinomycetes*. Moscow: Nauka (in Russian).

Gautam, S. P., Bundela, P. S., Pandey, A. K., Jamaluddin, Awasthi, M. K. & Sarsaiya, S. (2010). Optimization of the medium for the production of cellulase by the *Trichoderma* viride using submerged fermentation. *Int. J. Environ. Sci.*, *1*, 656-665.

Gomes, I., Sarkar, P. K., Rahman, S. R., Rahim, M. A. & Gomes, D. J. (2007). Production of cellulase from *Talaromyces emersonii* and evaluation of its application in eco-friendly functional finishing of jute-based fabrics. *Bangladesh. J. Microbiol.*, *24*, 109-114.

Gomes, N. C. M., Heuer, H. & Schonfeld, J. (2001). Bacterial diversity of the rhizosphere of maize (*Zea Mays*) grow in tropical soil studied by temperature gradient gel electrophoresis. *Plant. Soil.*, *232*, 167-180.

Gopinath, S. M., Shareef, I. & Ashalatha Ranjit, S. (2012). Isolation, screening and purification of cellulase from cellulase producing *Klebsiella variicola* RBEB3 (KF036184.1). *Int. J. Sci. Res.*, *3*, 1398-1403.

Gupta, C., Jain, P., Kumar, D., Dixit, A. K. & Jain, R. K. (2015). Production of cellulase enzyme from isolated fungus and its application as efficient refining aid for production of security paper. *Int. J. Appl. Microbiol. Biotechnol. Res.*, *3*, 11-19.

Hankin, L. & Anagnostakis, S. (1977). Solid media containing carboxymethylcellulose to detect Cx cellulase activity of micro-organisms. *J. Gen. Microbiol.*, *98*, 109-115.

Hayakawa, M., Yoshida, Y. & Iimura, Y. (2004). Selection of bioactive soil actinomycetes belonging to the *Streptomyces violaceusniger* phenotypic cluster. *J. Gen. Appl. Microbiol.*, *96*, 973–981.

Høiby, N., Krogh Johansen, H., Moser, C., Song, Z., Ciofu, O. & Kharazmi, A. (2001). *Pseudomonas aeruginosa* and the *in vitro* and *in vivo* biofilm mode of growth. *Microb. Infect.*, *3*, 23-35.

Ibrahim, A. S. S. & El-Diwany, A. (2007). Isolation and identification of new cellulases producing thermophilic bacteria from an Egyptian hot spring and some properties of the crude enzyme. *Aust. J. Basic. Appl. Sci.*, *1*, 473-478.

Ibrahim, M. F., Razak, M. N. A., Phang, L. Y., Hassan, M. A. & Abd-Aziz, S. (2013). Crude cellulase from oil palm empty fruit bunch by *Trichodermaasperellum* UPM1 and *Aspergillus fumigatus* UPM2 for fermentable sugars production *Appl. Biochem. Biotechnol.*, *170*, 1320-1335.

Irfan, M., Safdar, A., Syed, Q. & Nadeem, M. (2012). Isolation and screening of cellulolytic bacteria from soil and optimization of cellulase production and activity. *Turk. J. Biochem.*, *37*, 287-293.

Jang, H. D. & Chen, K. S. (2003). Production and characterization of thermostable cellulases from *Streptomyces* transformant T3-1. *World. J. Microbiol. Biotechnol.*, *19*, 263-278.

Jaradat, Z., Dawagreh, A., Ababneh, Q. & Saadoun, I. (2008). Influence of culture conditions on cellulase production by *Streptomyces* sp. (strain J2). *Jordan. J. Biol. Sci.*, *1*, 141-146.

Jeffrey, L. S. H. & Azrizal, M. R. (2007). Screening for cellulase activities in actinomycetes isolated from different locations of Peninsular Malaysia. *J. Trop. Agric. and Fd. Sc.*, *35*, 153-157.

Jiang, Y., Xie, C., Yang, G., Gong, X., Chen, X., Xu, L. & Bao, B. (2011). Cellulase-producing bacteria of *Aeromonas* are dominant and indigenous in the gut of *Ctenopharyngodon idellus* (Valenciennes). *Aquac. Res.*, *42*, 499-505.

Kanmani, R., Vijayabaskar, P. & Jayalakshmi, S. (2011). Sachaarification of banana-agro waste and clarification of apple juice by cellulase enzyme produced from *Bacillus pumilis*. *World. Appl. Sci. J.*, *12*, 2120-2128.

Karmakar, M. & Ray, R. R. (2011). Current trends in research and application of microbial cellulases. *Res. J. Microbiol.*, *6*, 41-53.

Kasana, R. C., Salwan, R., Dhar, H., Dutt, S. & Gulati, A. (2008). A rapid and easy method for the detection of microbial cellulases on agar plates using Gram's Iodine. *Curr. Microbiol.*, *57*, 503-507.

Kaur, J., Chadha, B. S., Kumar, B. A., Kaur, G. S. & Saini, H. S. (2007). Purification and characterization of β-glucosidase from *Melanocarpus* sp. MTCC 3922. *Electron. J. Biotechnol.*, *10*, 260-270.

Khianngam, S., Pootaeng-on, Y., Techakriengkrai, T. & Tanasupawat, S. (2014). Screening and identification of cellulase producing bacteria isolated from oil palm meal. *J. Appl. Pharm. Sci.*, *4*, 090-096.

Khokhar, I., Haider, M. S., Mushtaq, S. & Mukhtar, I. (2012). Isolation and screening of highly cellulolytic filamentous fungi. *J. Appl. Sci. Environ. Manage.*, *16*, 223-226.

Ko, C. H., Tsai, C. H., Lin, P. H., Chang, K. C., Tu, J., Wang, Y. N. & Yang, C. Y. (2010). Characterization and pulp refining activity of a *Paenibacillus campinasensis*cellulase expressed in *Escherichia coli*. *Bioresource. Technol.*, *101*, 7882-7888.

Kopečny, J., Hajer, J. & Mrázek, J. (2004). Detection of cellulolytic bacteria from the human colon. *Folia. Microbiol.*, *49*, 175-177.

Kuester, E. & Williams, S. T. (1964). Selection of media for isolation of *Streptomycetes*. *Nature*, *202*, 928-929.

Kuhad, R. C., Gupta, R. & Singh, A. (2011). Microbial cellulases and their industrial applications. *Enz. Res.*, (doi:10.4061/2011/280696).

Kumar, D., Ashfaque, M., Muthukumar, M., Singh, M. & Garg, N. (2012). Production and characterization of carboxymethyl cellulase from *Paenibacillus polymyxa* using mango peel as substrate. *J. Environ. Biol.*, *33*, 81-84.

Kumar, D., Yadav, K. K., Muthukumar, M., Kumar, S. & Garg, N. (2015). Mango juice clarification using purified fungal cellulase obtained from mango peel residue fermentation. *Acta. Hortic.*, *1066*, 213-216.

Kumar, U., Tapwal, A., Kalkal, P., Varghese, S. & Chandra, S. (2014). Isolation and screening of cellulase producing fungi from forest waste. *Int. J. Pharm. Biol. Arch.*, *5*, 56-59.

Ladeira, S. A., Cruz, E., Delatorre, A. B., Barbosa, J. B. & Martins, M. L. L. (2015). Cellulase production by thermophilic *Bacillus* sp. SMIA-2 and its detergent compatibility. *Electron. J. Biotechnol.*, *18*, 110-115.

Lakhundi, S., Siddiqui, R. & Khan, N. A. (2015). Cellulose degradation: a therapeutic strategy in the improved treatment of *Acanthamoeba* infections. *Parasites. Vectors.*, *8*, 23 (doi: 10.1186/s13071-015-0642-7).

Lee, Y. M., Lee, H., Kim, J. S., Lee, J., Ahn, B. J., Kim, G. H. & Kim, J. J. (2014). Optimization of medium components for β-glucosidase production in *Schizophyllum commune* KUC9397 and enzymatic hydrolysis of lignocellulosic biomass. *Bioresources.*, *9*, 4358-4368.

Liang, Y. L., Zhang, Z., Wu, M., Wu, Y. & Feng, J. X. (2014). Isolation, screening, and identification of cellulolytic bacteria from natural reserves in the subtropical region of China and optimization of cellulase production by *Paenibacillus terrae* ME27-1. *Biomed. Res. Int.* (doi:10.1155/2014/51249).

Li, D., Feng, L., Liu, K., Cheng, Y., Hou, N. & Li, C. (2016). Optimization of cold-active CMCase production by psychrotrophic *Sphingomonas* sp. FLX-7 from the cold region of China. *Cellulose.*, *23*, 1335-1347.

Li, X. (1997). *Streptomyces cellulolyticus* sp. nov., a new cellulolytic member of the genus *Streptomyces*. *Int. J. Syst. Bacteriol.*, *47*, 443-445.

Lin, Y., Wei, W., Yuanyuan, P. & Fuping, L. (2012). Screening and identification of cellulase-producing strain of *Fusarium oxysporum*. *Procedia. Environ. Sci.*, *12*, 1213-1219.

Lohlum, S. A., Forcados, E. G., Chuku, A., Agida, O. G. & Ozele, N. (2014). Corn cob as a feed component through fungal fermentation using *Aspergillus niger*. *CIBTech. J. Microbiol.*, *3*, 37-42.

Loiselle, M. & Anderson, K. W. (2003). The use of cellulase in inhibiting biofilm formation from organisms commonly found on medical implants. *Biofouling.*, *19*, 77-85.

Loliam, B., Morinaga, T. & Chaiyanan, S. (2013). Biocontrol of *Pythium aphanidermatum* by the cellulolytic actinomycetes *Streptomyces rubrolavendulae* S4. *ScienceAsia.*,*39*, 584-590.

Lotfi, A., Ghanbary, M. A. T., Ranjbar, G. A. & Asgharzadeh, A. (2010). Screening of some Zygomycetes for cellulase activity. *Afr. J. Biotechnol.*, *9*, 4211-4216.

Lugani, Y., Singla, R. & Sooch, B. S. (2015). Optimization of cellulase production from newly isolated *Bacillus* sp. Y3. *J. Bioprocess. Biotech.* 5, 264. (doi:10.4172/2155-9821.1000264).

Lynd, L. R., van Zyl, W. H., McBride, J. E. & Lasar, M. (2005). Consolidated bioprocessing of cellulosic biomass: an update. *Curr. Opin. Biotechnol.*, *16*, 577-583.

Maki, M. L., Broere, M., Leung, K. T. & Qin, W. (2011). Characterization of some efficient cellulase producing bacteria isolated from paper mill sludges and organic fertilizers. *Int. J. Biochem. Mol. Biol.*, *2*, 146-154.

Mandels, M. L., Hontz, L. & Nistrom, J. (1974). Enzymatic hydrolysis of waste cellulose. *Biotechnol. Bioeng.*, *16*, 471-493.

Mandels, M. & Reese, E. T. (1957). Induction of cellulase in *Trichoderma viride* as influenced by carbon sources and metals. *J. Bacteriol.*, *73*, 269–278.

Markovina, A., Pitt, J. I. & Hocking, A. D. (2005). Diversity of the *Trichocomaceae* in the Katandra Nature Reserve, Central Coast, NSW, Australia. *Mycol. Res.*, *109*, 964-973.

Menendez, E., Garcia-Fraile, P. & Rivas, R. (2015). Biotechnological applications of bacterial cellulases. *Bioeng.*, *2*, 163-182.

Mohagheghi, A., Grohmann, K., Himmel, M., Leighton, L. & Updegraff, D. M. (1986). Isolation and characterization of *Acidothermus cellulolyticus* gen. nov., sp. nov., a new genus of thermophilic, acidophilic, cellulolytic bacteria. *Int. J. Syst. Bacteriol.*, *36*, 435-443.

Mohanta, Y. K. (2014). Isolation of cellulose-degrading *actinomycetes* and evaluation of their cellulolytic potential. *Bioeng. Biosci.*, *2*, 1-5.

Moosavi-Nasab, M. & Majdi-Nasab, M. (2007). Cellulase production by *Trichoderma reesei* using sugar beet pulp. *Iran. Agr. Res.*, *25*, 107-116.

Moretti, M. M. S., Bocchini-Martins, D. A., Silva, R. D., Rodrigues, A., Sette, L. D. & Gomes, E. (2012). Selection of thermophillic and thermotolerant fungi for the production of cellulases and xylanases under solid-state fermentation. *Braz. J. Microbiol.*, 1062-1071.

Mrudula, S. & Murugammal, R. (2011). Production of cellulase by *Aspergillus niger* under submerged and solid state fermentation using coir waste as substrate. *Braz. J. Microbiol.*, *42*, 1119-1127.

Naveenkumar, K. J. & Thippeswamy, B. (2013). Isolation and screening of potential cellulolytic fungi from Areca nut husk waste. *Int. J. Curr. Sci.*, *8*, 125-132.

Nisha, P. (2015). Cellulase production optimization using cellulolytic bacteria. *Int. J. Pharmaceut. Chem. Biol. Sci.*, *5*, 262-266.

Ogbonna, A. I., Nwadiaro, P. O., Chuku, A., Ogbonna, C. I. C. & Onwuliri, F. C. (2015). Cellulase induction in three *Aspergillus* species isolated from *Artemisia annua* L. plantation soil using different cellulose substrates. *J. Appl. Biotechnol.*, *3*, 63-74.

Padmavathi, T., Nandy, V. & Agarwal, P. (2012). Optimization of the medium for the production of cellulases by *Aspergillus terreus* and *Mucor plumbeus*. *Euro. J. Exp. Bio.*, *2*, 1161-1170.

Patagundi, B. I., Shivasharan, C. T. & Kaliwal, B. B. (2014). Isolation and characterization of cellulase producing bacteria from soil. *Int. J. Curr. Microbiol. App. Sci.*, *3*, 59-69.

Pečiulytė, A. (2007). Isolation of cellulolytic fungi from waste paper gradual recycling materials. *EKOLOGIJA.*, *53*, 11-18.

Picart, P., Diaz, P. & Pastor, F. I. J. (2007). Cellulases from two *Penicillium* sp. strains isolated from subtropical forest soil: production and characterization. *Lett. Appl. Microbiol.*, *45*, 108-113.

Pirzadah, T., Garg, S., Singh, J., Vyas, A., Kumar, M., Gaur, N., Bala, M., Rehman, R., Varma, A., Kumar, V. & Kumar, M. (2014). Characterization of Actinomycetes and *Trichoderma* spp. for cellulase production utilizing crude substrates by response surface methodology. *SpringerPlus.*, *3*, 622.

Prasad, M. P. & Sethi, R. (2013). Optimization of cellulase production from a novel bacterial isolate *Mesorhizobium* sp. from marine source. *J. Enzyme. Res.*, *4*, 39-45.

Prasad, P., Singh, T. & Bedi, S. (2013). Characterization of the cellulolytic enzyme produced by *Streptomyces griseorubens* (Accession No. AB184139) isolated from Indian soil. *J. King. Saud. Univ. Sci.*, *25*, 245-250.

Prasad, P., Singh, T. & Bedi, S. (2014). Enzymatic screening and characterization of cellulolytic actinomycetes isolated from soil samples collected from Patna district in Bihar, India. *Int. J. Curr. Res. Aca. Rev.* 2, 60-71.

Pratima, G., Samant, K. & Sahu, A. (2012). Isolation of cellulose-degrading bacteria and determination of their cellulolytic potential. *Int. J. Microbiol.* Article ID 578925, 1-5.

Premalatha, N., Gopal, N. O., Jose, P. A., Anandham, R. & Kwon, S. W. (2015). Optimization of cellulase production by *Enhydrobacter* sp. ACCA2 and its application in biomass saccharification. *Front. Microbiol.*, *6*, 1046. (doi:10.3389/fmicb.2015.01046).

Rajasekharan, S. K. & Ramesh, S. (2013). Cellulase inhibits *Burkholderiacepacis* biofilms on diverse prosthetic materials. *Polish. J. Microbiol.*, *62*. 327-330.

Rathna, G. S., Saranya, R. & Kalaiselvam, M. (2014). Bioethanol from sawdust using cellulase hydrolysis of *Aspergillus ochraceus* and fermentation by *Saccharomyces cerevisiae*. *Int. J. Curr. Microbiol App. Sci.*, *3*, 733-742.

Rathnan, R. K. & Ambili, M. (2011). Cellulase enzyme production by *Streptomyces* sp. using fruit waste as substrate. *Aust. J. Basic. Appl. Sci.*, *5*, 1114-1118.

Ravindran, C., Naveenan, T. & Varatharajan, G. (2010). Optimisation of alkaline cellulase production from marine derived fungi, *Chaetomium* sp., using agricultural and industrial wastes as substrates. *Bot. Mar.*, *53*, 275-282.

Robert, C. & Bernalier-Donadille, A. (2003). The cellulolytic microflora of the human colon: evidence of microcrystalline cellulose-degrading bacteria in methane-excreting subjects. *FEMS. Microbiol. Ecol.*, *46*, 81-89.

Sadhu, S. & Maiti, T. K. (2013). Cellulase production by bacteria: A review. *Brit. Microbiol. Res. J.*, *3*, 235-258.

Safdar, A., Irfan, M., Nadeem, M. & Syed, Q. (2013). Carboxymethyl cellulase production from newly isolated *Cellulomonas* sp. in submerged fermentation. *Hacettepe. J. Biol. Chem.*, *41*, 179-185.

Sambasivarao, S. V., Granum, D. M., Wang, H. & Maupin, C. M. (2014). Identifying the enzymatic mode of action for cellulase enzymes by means of docking calculations and a machine learning algorithm. *AIMS. Mol. Sci.*, *1*, 59-80.

Sami, A. J., Awais, M. & Shakoori, A. R. (2008). Preliminary studies on the production of endo-1,4-β-D-glucanases activity produced by *Enterobacter cloacae*. *Afr. J. Biotechnol.*, *7*, 1318-1322.

Selvam, K., Vishnupriya, B. & Bose, V. S. C. (2011). Screening and quantification of marine actinomycetes producing industrial enzymes amylase, cellulase and lipase from south coast of India. *Int. J. Pharm. Biol. Arch.*, *2*, 1481-1487.

Shankar, T. & Isaiarasu, L. (2011). Cellulase production by *Bacillus pumilus* EWBCM1 under varying cultural conditions. *Middle-East J. Sci. Res.*, *8*, 40-45.

Shirling, E. B. & Gottlieb, D. (1966). Methods for characterization of *Streptomyces* species. *Int. J. Syst. Bacteriol.*, *16*, 313-340.

Shweta, A. (2012). Cellulases of bacterial origin and their applications: a review. *Int. J. Sci. Res.*, *3*, 1652-1655.

Singh, M., Khanna, S. & Prakash, N. T. (2010). Influence of cellulolytic bacteria augmentation on organic carbon and available phosphorus in sandy loam soil under cultivation. *J. Agr. Sci.*, *2*, 137-145.

Singh, S., Gadekar, V. & Khan, F. (2015). Preliminary characterization of cellulases from actinomycetes *Streptomyces cinnabarinus* and *Streptomyces globosus* isolated from wastewater effluents of pulp industries. *Int. J. Innov. Biol. Res.*, *4*, 4-8.

Siripornadulsil, S., Poomai, N. & Siripornadulsil, W. (2014). Application of *Acinetobacter* sp. KKU44 for cellulase production from agricultural waste. *Int. Scholarly. Sci. Res. Innov.*, *8*, 373-375.

Sleat, R., Mah, R. A. & Robinson, R. (1984). Isolation and characterization of an anaerobic, cellulolytic bacterium, *Clostridium cellulovorans* sp. nov. *Appl. Environ. Microbiol.*, *48*, 88-93.

Sohail, M., Ahmad, A. & Khan, S. A. (2011). Production of cellulases from *Alternaria* sp. MS28 and their partial characterization. *Pak. J. Bot.*, *43*, 3001-3006.

Sohail, M., Ahmad, A. & Khan, S. A. (2016). Production of cellulase from *Aspergillus terreus* MS105 on crude and commercially purified substrates. *3 Biotech.*, *6*, 103 (doi: 10.1007/s13205-016-0420-z).

Song, J. M. & Wei, D. Z. (2010). Production and characterization of cellulases and xylanases of*Cellulosimicrobium cellulans* grown in pretreated and extracted bagasse and minimal nutrient medium M9. *Biomass. Bioenergy.*, *34*, 1930-1934.

Sreedevi, S., Sajith, S. & Benjamin, S. (2013). Cellulase producing bacteria from the wood-yards on Kallai river bank. *Adv. Microbiol.*, *3*, 326-332.

Sri Lakshmi, A. & Narasimha, G. (2012). Production of cellulases by fungal cultures isolated from forest litter soil. *Ann. For. Res.*, *55*, 85-92.

Stewart, J. C., Stewart, C. S. & Heptinstall, J. (1982). The use of tritiated cellulose in screening for cellulolytic microorganisms. *Biotechnol. Lett.*, *4*, 459-464.

Sukumaran, R. K., Singhania, R. R. & Pandey, A. (2005). Microbial cellulase-Production, applications and challenges. *J. Sci. Ind. Res.*, *64*, 832-844.

Swaroopa Rani, D., Thirumale, S. & Nand, K. (2004). Production of cellulase by *Clostridium papyrosolvens* CFR-703. *World. J. Microbiol. Biotechnol.*, *20*, 629-632.

Teather, R. M. & Wood, P. J. (1982). Use of congo red-polysaccharide interactions in enumeration and characterization of cellulolytic bacteria in the bovine rumen. *Appl. Environ. Microbiol.*, *43*, 777-780.

Tejada, M., Gonzalez, J. L., García-Martínez, A. M. & Parrado, J. (2008). Application of a green manure and green manure composted with beet vinasse on soil restoration: effects on soil properties. *Bioresource. Technol.*, *99*, 4949-4957.

Wang, A., Gao, L., Ren, N., Xu, J., Liu, C., Cao, G., Yu, H., Liu, W., Hemme, C. L., He, Z. & Zhou, J. (2011). Isolation and characterization of *Shigella flexneri* G3, capable of effective cellulosic saccharification under mesophilic conditions. *Appl. Environ. Microbiol.*, *77*, 517-523.

Wang, D., Qu, Y. B. & Gao, P. J. (1995). The mechanism of increasing cellulase biosynthesis rate in *Trichoderma* by L-sorbose. *Acta. Mycol. Sinica.*, *14*, 143-147.

Wolf, F. T. (1952). The production of indole acetic acid by *Ustilago zeae*, and its possible significance in tumor formation. Proceedings of the National Academy of Sciences of the United States of America., *38*, 106-111.

Wongputtisin, P., Khanongnuch, C., Kongbuntad, W., Niamsup, P., Lumyong, S. & Sarkar, P. K. (2014). Use of *Bacillus subtillis* isolates from Tua-nao towards nutritional improvement of soya bean hull for monogastric feed application. *Lett. Appl. Microbiol., 59*, 328-333.

Wood, T. M. & Bhat, K. M. (1988). Methods for measuring cellulase activities. *Methods. Enzymol., 160*, 87-117.

Xin, F., Zhang, H. & Wong, W. (2013). Bioethanol production from horticultural waste using crude fungal enzyme mixtures produced by solid state fermentation. *Bioenergy. Res., 6*, 1030-1037.

Yang, W., Meng, F., Peng, J., Han, P., Fang, F., Ma, L. & Cao, B. (2014). Isolation and identification of a cellulolytic bacterium from the Tibetan pig's intestine and investigation of its cellulase production. *Elect. J. Biotechnol., 17*, 262-267.

Yin, L. J., Huang, P. S. & Lin, H. H. (2010). Isolation of cellulase-producing bacteria and characterization of the cellulase from the isolated bacterium *Cellulomonas*sp. YJ5. *J. Agric. Food Chem.,58*, 9833-9837.

Zucker, M. & Hankin, L. (1970). Regulation of pectate lyase synthesis in *Pseudomonas fluorescens* and *Erwinia caratovora. J. Bacteriol., 104*, 13-18.

In: Cellulase
Editor: Marguerite Stevenson

ISBN: 978-1-63485-788-8

Chapter 2

MICROBIAL CELLULASE: PRODUCTION AND APPLICATIONS IN ENZYMATIC TREATMENT OF BIOSOLIDS

A. A. Al-Gheethi[1,*], A. N. Efaq[2,3], R. M. S. R. Mohamed[1], J. D. Bala[4] and M. K. Amir Hashim[1]

[1]Micro-pollution Research Centre (MPRC),
Department of Water and Environmental Engineering,
Faculty of Civil and Environmental Engineering,
Universiti Tun Hussein Onn Malaysia, Batu Pahat, Johor, Malaysia
[2]Environmental Technology Division, School of Industrial Technology,
Universiti Sains Malaysia (USM), Penang, Malaysia
[3]Applied Microbiology Department, Faculty of Applied Sciences,
Taiz University, Taiz, Yemen
[4]Department of Microbiology, Federal University of Technology,
Minna, Abuja, Minna, Nigeria

ABSTRACT

One of the main objectives in the treatment of biosolids is to reduce the cellulolytic materials which represent 40 to 50% of the total dry weight. Cellulose degradation is initiated primarily by micro-organism (fungi, bacteria and protozoa) with the aid of extracellular enzymes. The

* Email: alisaeed@uthm.edu.my; adelalghithi@gmail.com.

cellulase(s) is a multiple enzyme system consisting of endoglucanase, exoglucanase and β-glucosidase which act as a synergistic effect for the degradation of complex cellulose to glucose. Cellulase enzyme has applied for several industrial activities including production of animal feed, and wine, formulation of detergents, juice clarification, as well as in the treatment of domestic and industrial wastes. The enzymatic treatment of biosolids is named white technology which has raised recently due to the efficiency and absence of toxic by-products. In this chapter the factors affecting production of cellulase enzyme by microorganisms, the potential of indigenous microorganisms for the bio-augmentation process of biosolids and occurrence of catabolite repression phenomenon of cellulase production during the enzymatic degradation of biosolids will be discussed.

Keywords: cellulase, microorganism, biosolids, catabolite repression

INTRODUCTION

Cellulose represent the main component in the biosolids generated from sewage treatment plants (STPs). It represent between 40-50% of the dry weight of biosolids (Wang et al., 1994). Therefore, the reduction for the total mass of biosolids is the major concern of the treatment processes. Currently the biosolids generated from STPs are incinerated or disposed in to the landfill. However, these practice are unacceptable due to high risk associated with the incineration process such as production of secondary by-products which including dioxins and furans (Efaq et al., 2015). The direct disposal of biosolids into the landfill might contribute significantly in the distribution of pathogenic microorganisms to the environment and then transmit via food chain into the human. Biosolids have high concentrations of pathogenic bacteria, viruses, protozoa and fungi (Abdel-Monem et al., 2008a). These organisms have the ability to survive in the environment for several months and sometime for years (Abdel-Monem et al., 2008a). In some developing countries biosolids generated from STPs are used as soil fertilizers, these practices accelerate the distribution and transmission of pathogens into the human and animals (Al-Gheethi et al., 2015a).

Several technologies have been rised during the last decades, which aimed to reduce the health risk and increase the total dry mass of biosolids. These technologies included thickening, stabilization, conditioning and de-watering processes (Abdel-Monem et al., 2008b). However, they could not contribute in the reduction of the biosolids volume. The new strategies are to reuse the biosolids as biofuel resources. Cellulose represent one of the important resources for biofuel production, however, it has to subject for the enzymatic hydrolysis using cellulase enzyme in which the polysaccharides are degraded into glucose that can be used as substrate for fuel ethanol production (Salmon et al., 2014). The degradation of cellulolytic materials by cellulase(s) enzymes produced from numerous microorganisms is very important in several agricultural and biosolids treatment processes (Hamer, 2003; Angenent et al., 2004; Schloss et al., 2005). Since, the cellulose is very stable with a half-life of 5-8 million years (Wolfenden & Snider, 2001).

The enzymatic treatment process is a technology falls between the biological and physicochemical processes (Karam & Nicell, 1997). The Bio-catalysis technology which is based on utilization of enzymes is known as white biotechnology and fully participates in the "green chemistry" concept (Alcalde et al., 2006). It has introduced in the 90s by Sheldon & van Rantwijk (2004). The enzymatic technology has several advantages including: the high potential to degrade complex compounds, effective under a wide range of temperature, salinity and pH (Karam & Nicell, 1997). It is eco-friendly process which means there is not toxics are produced, conversely it remediate a wide range of compounds unfriendly to the environment.

In this chapter the microbial production of cellulase enzyme, mechanism of action, the application of cellulase for the reduction of biosolids volume and the challenges are discussed.

1. Microbial Production of Cellulase Enzyme

The production of cellulase among different microorganisms have been reported in literature. Several anaerobic bacteria degrade cellulose in anaerobic digestion (Chynoweth & Pratap, 1996). The aerobic microorganisms have high potential to produce copious amounts of free cellulases which act synergistically to degrade cellulose (Blouzard et al., 2007; Mingardon et al., 2007). *Cellulomonas sp. is the most common bacteria which has the ability to produce high concentrations of cellulase. Other bacteria including Bacillus sp. and Pseudomonas sp. Among several species of fungi Aspergillus spp. have high potential to degrade the cellulolytic materials and produce cellulase enzyme.*

Al-Gheethi (2015) found that Sporosarcina pasteurii, Bacillus megaterium, Staphylococcus xylosus, B. subtilis and Burkholderia *cepacia* obtained from wastewater and biosolids have exhibited high productivity of cellulase enzymes. Bala et al., (2015) isolated *Micrococcus luteus, S. maltophilia, B. cereus, K. pneumoniae, B. subtilis, Aspergillus fumigatus, A. nomius, A. niger, Meyerozyma guilliermondii* from POME and revealed that these species exhibited high activity for the production of cellulase. Efaq et al., (2016a) found that *A. niger, A. fumigatus, A. tubingensis, A. terreus var. terreus, P. simplicissimum* and *P. waksmanii* obtained from clinical wastes have a cellulase activity on the production medium.

The first screening for the production of cellulase enzyme among the microorganisms is conducted by using the pate technique which has introduced by Hankin & Anagnostakis (1975). The technique is depend on the subculture of the microorganism on the culture medium containing cellulase substrate. The most common substrate used for the detection of cellulase enzyme is Carboxymethyl cellulose (CMC). A pure culture of the microorganism is streaked on the CMC medium and then incubate at 28°C for fungi and 35-37°C for bacteria. After the incubation period for 3-5 days, the CMC plates are flooded with 1% (w/v) congo red for 20 min and then the dye is decanted and the plates are again flooded with one molar (1 M) NaCl for another 20-30 min (Singh et al., 2013; Bala et al., 2015; Efaq et al., 2016b). The cellulase production is detected by the presence of a pale orange to clear zone around the microorganism colony which indicate to cellulose hydrolysis (Sadhu & Maiti, 2013). The occurrence of clearing zone around the cellulolytic bacteria or fungi colony is due to colour change of Congo red which take place as a result of degradation of cellulose into glucose and routinely metabolize glucose to organics acids. This process lead to decrease the pH to be in acidic condition and then Congo red change from red to pale orange (Yoon et al., 2007). Another reagent for the detection of cellulase enzyme might be conducted by using iodine $ZnCl_2$ solution (3% $ZnCl_2$ is added to gram's iodine solution), where the appearance of clearing zone around the colony mean the production of enzyme is positive (Al-Gheethi, 2015).

In order to confirm the presence of extracellular cellulase, the microorganism is cultured in a CMC-broth medium for 5-7 days. The cells are separated by the centrifugation 4000-6000 rpm for 10 min to obtain the supernatant. The detection of cellulase in the supernatant is carried out by using CMC cup-plate clearing zone (C.C.Z) assay. In this assay 2% (w/v) CMC is dissolved in citrate buffers (pH 7) for bacteria or phosphate buffer for fungi and supplement with 1.5% agar for solidification. After sterilization, equal amounts of assay medium (20 ml) is poured in sterilized petri dishes (12 cm in diameter). Cups (10 mm in diameter) is made in each plate using a sterile cork borer. Equal amounts of the enzyme solution (cultural supernatant) are put into each cup. Plates with cups containing enzyme solutions are incubated at 37°C for 24 h, then the surface of the plates were floated with the reagent (Congo red or $ZnCl_2$ solution. Diameters of blue zones are measured (Al-Gheethi, 2015).

The concentration of cellulase enzyme is determined by using spectrophotometer by using the dinitrosalicylic (DNS) method according to Miller (1959). *In this method* glucose is used as the standard. Measurements are made in a spectrophotometer at 540 nm wavelength in the presence of the blank which is containing heat-inactivated post-culture liquids (boiled for 5 min).

Cellulases enzymes are complex enzyme consist of endoglucanases (EC3.2.1.4), exoglucanases, including D-cellodextrinases (EC3.2.1.74) and cellobiohydrolases (EC 3.2.1.91) and (3) b-glucosidases (EC 3.2.1.21) (Lynd et al., 2002; Zhang & Lynd, 2004). Endoglucanase hydrolyze accessible intra-molecular β-1-4 glucosidic bonds of cellulose chain ends, exoglucanases processively cleave cellulose soluble cellobiose or glucose and β-glucosidase hydrolyze cellobiose to glucose in order to cellobiose inhibition (Krishna, 1999; Zhang et al., 2006).

2. Enzymatic Treatment of Biosolids

The main objective of wastewater treatment plants is to degrade the total suspended solids (TSS) which represent the cellulolytic materials in wastewaters (Wu et al., 2010). The primary and secondary process contribute significantly in the reduction of TSS in the effluents. However, these concentrations still high in the biosolids which remains a problem to be resolved further (AbdulKarim et al., 2011). The use of microbial enzymes as an advance technology for biosolids might offers an alternative solution to reduce the chemical oxygen demand (COD), biological oxygen demand (BOD) and total suspended solids (TSS) parameters (Alam et al., 2009).

Microbial cell degrade the organic compounds by using two types of enzymes which are endo-enzymes and exo-enzymes. Endo-enzymes are produced within the cell and used for the metabolic reactions of soluble substrates. In this case the substrates with low molecules weight are transported across the cell membrane via the transport systems of the microorganism into the cytoplasm. Exo-enzymes are synthesized inside the microbial cell and then release to the surrounding medium to breakdown the high molecular compounds which could not transported across the cell membrane of the cell (Gerardi, 2003). The cellulose has high molecular weight with about 570,000 units of glucose. Therefore, the microbial cell produce cellulase enzyme as a hydrolysis enzyme to degrade it into small units with low molecular weight like glucose (180 MW) to facility the transportation process through the cell membrane of the microorganism.

The simplest method for the enzymatic treatment of biosolids is by introducing of the microbial cell that produce cellulase into the biosolids reactor. However, it has to consider several factors which affecting microbial cellulase production in biosolids such as initial concentrations of the microorganisms, temperature, pH value, carbon and nitrogen sources as well as heavy metals concentrations in biosolids. Besides, the potential of the introduced microorganism has to survive and compete with the indigenous organisms in the biosolids. In terms of initial concentrations of inoculated microorganism, it has to mention that the concentrations rely on the age of microorganism for example for the bacterial cells it supposed to be between 24 to 48 hrs, while for the fungi it has to be between 72 and 96 hrs. At this age the microbial cells still have high activity for the production of the enzymes, because it will be in the stationary phase of the growth curve. Krishna (1999) revealed that 15% (v/w) of *B. subtilis* (24 hrs) produced maximum concentrations of the enzymes in the culture medium. Most of the researchers used the bacteria such as *B. pumilus, Streptomyces omiyaensis, Psudomonas fluorescens* within 24 hrs of the age as inoculum for the production of cellulase in the medium with initial concentrations ranged from a loop of the pure culture to 15% of the production medium without need to detect the numbers of the bacterial cells concentrations (Kotchoni et al., 2003; Bakare et al., 2005; Al-Gheethi 2015). Moreover, Al-Gheethi & Norli (2014) indicated that 10^6 cell/mL is the best initial concentrations for the enzymatic degradation of pharmaceutical restudies in the wastewater. It has to indicate that the initial concentrations of inoculated microbial cells would not effect directly on the amount of the cellulase production, since the microorganism will grow and multiply during the treatment process. Therefore, the microbial concentrations during the enzymatic treatment process of biosolids can be calculated according to the following equations (Davis and Cornwell 2013).

$$P = P_0 + 2P_0 + 2 \times 2P_0 + 2 \times 2 \times 2P_0 + \cdots \quad (1)$$

where P is the population, P_0 is the initial concentrations, and due to the multiplication of bacterial cells by Logatherm numbers the total population is calculated as following;

$$P = P_0(2)^n$$

Where n is the generation number which is depend on the generation time and different form one microorganism to others and can calculated as following:

$$n = \frac{treatment\ period}{generation\ time} \qquad (2)$$

The cellulolytic microbial growth during the treatment process is depend on the concentrations of cellulose in the biosolids as substrate. Therefore, in order to determine the microbial growth the Monod Equation can be used.

$$\mu = \mu_{max}\left[\frac{S}{S+Ks}\right] \qquad (3)$$

where, μ_{max} = maximum growth rate, t^{-1}, S = concentration of cellulose in the biosolids (in term of BOD), mg/L, K_s = half saturation constant, mg/L which can be calculated from the growth curve slop of substrate concentrations and microbial concentrations.

The growth rate of microorganisms can be determined by using the following question

$$\frac{dX}{dt} = \frac{\mu_m SX}{K_s + S} - k_d X \qquad (4)$$

where, dX/dt is growth rate of biomass, X is concentration of biomass, k_d is endogenous decay (natural die-off of microorganism during the treatment process).

The degradation of cellulolytic materials in the biosolids (in term of BOD) is calculated based on the microbial growth rate and decimal fraction of cellulolytic material mass utilized by the microorganism (Y) as following:

$$-\frac{dS}{dt}=\frac{1}{Y}\frac{\mu_m SX}{K_s+S}\ldots \quad (5)$$

$$Y\ (yield\ coefficient)=\frac{microorgasim\ concentration}{BOD\ reduced} \quad (6)$$

The effect of temperature on the production of cellulase enzyme depend on the microorganism species (bacteria/fungi). Most environmental microorganisms grow with the ambient temperature (25-30°C). However, the microbial cells have a wide range temperature for the production of cellulase enzyme. Bacillus spp. and Aspergillus spp. exhibited high potential to produce the enzyme within the range between 30°C to 60°C (Mawadza & Zvauya, 1996; Krishna, 1999; Immanuel et al., 2006; Ikram-ul-Haq et al., 2006; Al-Gheethi 2015). Moreover, the selection of inoculated microorganism species represent the main point in the degradation of cellulolytic materials in biosolids. The utilization an environmental microbe strain might exhibit high effective in the production of the enzyme and thus the treatment process can be conducted at the mesophilic temperature. In most studies the microorganism used are obtained from the biosolids itself because it has already acclimatized naturally to grow and multiply in this medium. This point is very important in the enzymatic treatment of the biosolids as will be discussed below.

The maximum production of cellulase enzymes might take place after 4 hrs to three days of the inoculation period for bacteria and might extend to seven days for fungi depend on the microorganism source (Kawai et al., 1988; Ekperigin, 2007). Microorganism needs time to adapt in the new media, the period of the adaption process depend on the microorganism source. Therefore, if the microorganism has obtained from the biosolids, the adaption period will take short time, then the enzymatic degradation of cellulose in biosolids will be fast.

pH value of biosolids is around pH 7, in contrast, microorganism have a wide pH range for the production of cellulase enzymes. Fungi are acidophilic, *Trichoderma reesi* produced the maximum enzyme between pH 2.8 and pH 3.5 (Sternberg & Mandels, 1979), while bacteria such as *Cellulomonas* spp., *Bacillus* spp. and *Micrococcus* spp. produce the enzyme at natural pH (Mawadza & Zvauya, 1996; Immanuel et al., 2006). However, some species of bacteria and fungi have the potential to produce the enzyme at alkaline conditions more than pH 9 (Hakamada et al., 1997). The wide range of pH required for the maximum production of cellulase enzyme facility the enzymatic treatment of biosolids where no adjustment of pH value by using chemical additives are required.

The type and concentration of cellulosic material and carbon source is critical for the production of cellulase enzyme by introduced microorganism in the biosolids treatments. Biosolids contains 40-50% cellulose, 12% hemicellulose and 10-15% lignin by dry weight (Wang et al., 1994). These substrates can be used as a carbon source for the cellulolytic organisms. Al-Gheethi (2015) examined CMC, cellulose powder, cotton, filter paper, tobacco leaves, sawdust and bagasse as a carbon sources for *B. megaterium* which is used for the production of cellulase enzymes in the production medium. The study revealed that CMC was the best for the maximum production of cellulase followed by cellulose powder. The lowest cellulase production was recorded with cotton and sawdust. CMC was also reported as the best carbon source for the production of cellulase enzyme among fungi such as *A. niger, T. viride, Paenibacillus* sp. (Ikram-ul-Haq et al., 2006; Emtiazi et al., 2007). Efaq et al., (2016a) used CMC as a substrate for the bioassay of cellulase enzymes among *A. niger, A. fumigatus, A. tubingensis, A. terreus var. terreus, P. simplicissimum* and *P. waksmanii.* In the investigation for the effect of carbohydrate on the cellulase production, Al-Gheethi (2015) found that the mannose, ribose, fructose and xylose have induced the production of cellulase by *B. megaterium.*

The nitrogen source in the medium is one of the most important factors, which greatly affect cellulase production by microorganisms. Biosolids are rich with organic nitrogen sources such as ammonium and urea which play an important role in the enhancement of cellulase production by the introduced microorganism (Chan & Au 1987). Jin & Toda (1989) demonstrated that the increase of urea concentration from 2 to 6 g L^{-1} and reduction of yeast extract from 6 to 4 g L^{-1} in the medium improved endocellulase production by Clostridium thermocopiae. Al-Gheethi & Abdul-Monem (2014) revealed that the organic nitrogen source has improved the production of cellulase enzyme by *B. megaterium* more than the inorganic source because the organic nitrogen compounds stimulated higher growth and cellulase production than inorganic compounds.

The composition of biosolids is the main challenges in the enzymatic treatment process, since the biosolids have many of heavy metals elements which might inhibit the microbial growth and then the production of cellulase enzymes. There are few studies which conducted on the production of cellulase enzyme in the biosolids media. Al-Gheethi (2015) examined the efficiency of bacteria to produce cellulase enzyme in the biosolids medium which contains different concentrations of heavy metals such as zinc (Zn), copper (Cu) and nickel (Ni). The study examined four types of biosolids collected from four different STPs and inoculated with *B. megaterium*. The efficiency of bacteria to produce of cellulase was conducted without chemical additives or pH adjustments. The study found that production of cellulase enzyme was high in the biosolids with low concentrations of Ni^{2+} even when the concentration of Cu^{2+} and Zn^{2+} ions were high. It was due to the toxicity of Ni^{2+} ions in comparison to Cu^{2+} and Zn^{2+} at the high concentration. Some of heavy metals such as Zn, Cu and Ni at the low concentration play role as cofactor for many of enzymes and improve the enzymatic reaction (Nies, 1999). However, at high concentrations some of these metals such as Ni^{2+} becomes more toxicity than Cu^{2+} and Zn^{2+}. The toxicity of heavy metals is depend on the transport system of microbial cells. For example, the microbial cell take up of $Zn^{2+}$and Cu^{2+} is by specific transport system (Fath & Kolter 1993; Fagan & Saier 1994), while Ni^{2+} is accumulating by the fast and unspecific system (Tao et al., 1995).

On the other hand, the ability of microbial cell to resist heavy metals might improve the over production of cellulase. For example, Al-Gheethi (2015) revealed that *B. megaterium* has the potential to produce detectable amount of cellulase in biosolids with Ni^{2+} ions was due to that *B. megaterium* was acclimatized to tolerate nickel ions before the inoculation to the biosolids medium. Therefore, one of the consideration for the utilization of microbial cells for the enzymatic treatment of biosolids is the selection process for the most potent microorganism.

The adding selected indigenous microbial strains into the biosolids which called bioaugmentation process, might be efficient in degrading cellulolytic materials and can improve the catabolism of specific compounds. It is indeed a promising technique to solve practical problems in wastewater treatment plants, and to enhance removal efficiency (Herrero & Stuckey 2015).

The introduced microorganism into the enzymatic treatment of biosolids should be subjected for several adaptation process in order to get high degradation process of cellulose. Several authors have applied the adaption strategy for the microorganism used in the bioagumentaion process to achieve high efficiency in the biodegradation of organic pollutant in the wastewater. Among bacterial species *P. aeruginosa, B. flexus, B. subtilis, E. homiense* and *S. aureus* (Sivaprakasam et al., 2008; Li et al., 2010; Al-Gheethi & Norli 2014; Adel et al., 2015; Al-Gheethi et al., 2015; Al-Gheethi et al., 2016a).

The considerations in the bioagumentaion process of enzymatic treatment of biosolids included the insufficiency of substrate, competition with indigenous organisms, pH, moisture, nutrient, osmotic factor. Hence, the choice of organisms is particularly important because of differences in their ability for growth and survival under extreme conditions and affinity for pollutants (Morikawa, 2006). The adaption process depend on the natural selection process which rely on the acclimatization and the competition process for resources (Davies & Davies, 2010; Radwan & Babik, 2012; Al-Gheethi et al., 2016b).

Recently, some researchers trended to use hybrid technology for the biotechnology, such as using the genetic engineering technology to improve the efficiency of introduced microorganisms (Wang & Chen, 2009). However, the limitation in the using genetically modified organisms (GMOs) are their ability to generate persistent strains which might give adverse effect for the natural ecosystems. Therefore, they have been considered a threat to the environment and human health (Gilsdorf & Zilinskas, 2005; Prescott et al., 2005; Prakash et al., 2011). Hence, the natural adaption process is the best to get more effective organism in the bioaugmentation process. This process has been used to the biodegradation of cellulose in biosolids and pharmaceutical residues in the wastewater and exhibited a good efficiency (Al-Gheethi 2015; Al-Gheethi et al., 2016a).

3. Catabolite Repression of Enzyme Production

One of the main challenges in the enzymatic treatment of biosolids lie in the Carbon catabolite repression (CCR) for the production of cellulase by the inoculated microorganisms. This process lead to inhibit the production of cellulase due to high concentrations of glucose in the medium. CCR is a key regulatory system available in the microorganism cell to ensure preferential utilization of energy-efficient carbon sources (Vinuselvi et al., 2012). Cellulase enzyme are produced by the microorganism as genetically or inducible. In the microorganism which produce the enzyme genetically the high level of glucose would not lead to inhibit the enzyme production. In contrast, in the inducible microorganism the presence of glucose in the surrounding medium might reduce the production of cellulase enzyme. Therefore, in the selection process of microorganism the nature of cellulase enzyme production as genetic or inducible should be tested. The test can be conducted by the sub-culturing of microbe in two production medium, one of the medium containing cellulase substrate such as CMC, while the other containing glucose and then determine the amount of the cellulase enzymes in both culture medium. The inducible organism produce less cellulase enzyme in the presence of glucose in comparison to that in the presence of inducer substrate. Al- Gheethi, (2015) has reported that *S. pasteurii, B. subtilis* and *S. Xylosus* lose the ability to produce cellulase in the presence of glucose, while B. megaterium has produced the enzyme genetically.

Glucose inhibit the production of cellulase enzyme in the inducible organism because it easily across through cell membrane and stop the gene expression of enzyme production. However, it has to mention that the CCR occur after several days of the enzymatic treatment process. Bala (2016) found that the CCR exhibited after 50 days of enzymatic treatment of POME by bacteria and fungi. Allcock & Woods (1981) and Al-Gheeth (2015) has reported that microbial strains can continually produce hydrolysis enzymes if it does not undergo catabolite repression to degrade its substrate (cellulose) and obtain the end products (glucose) which can be detectable as long as the degrading microbe produce the enzymes. Even though microbes can utilize glucose for their survival (energy) it will be very less as compare to the production of the glucose in the medium.

Catabolite repression inhibit or repress the production of hydrolytic enzymes of an organisms in the presence of glucose which will cause decrease in the reduction of the parameters since the enzymes involve in the degradation process is inhibited. The CCR can also occur with others enzymes such as lipase, Boekema et al., (2007) has reported that microbial cell does not produce lipase in the medium containing glucose suggesting that lipase gene expression in the microbial cell was prone to catabolite repression. The presence of glucose inhibited or repressed the production of lipase in the medium and hence the degradative activity of the enzymes was inhibited.

CONCLUSION

Cellulase enzyme is one of the most important hydrolysis enzymes which might play an important role in the enzymatic treatment of biosolids. However, the considerations with the application of the enzymatic technology lie in the selection process for the most potent microorganism which is high effective in different types of biosolids compositions.

ACKNOWLEDGMENTS

The authors gratefully acknowledge Ministry of Higher Education of Malaysia for the research project financial support under fundamental research grant scheme (FRGS) vot No. 1453 and prototype research grant scheme (PRGS) vot G004.

REFERENCES

Abdel-Monem, M. O., Al-Zubeiry, A. H. & Al-Gheethi A. A. (2008). Elimination of *Salmonella* and faecal indicator bacteria in sewage sludge by quick lime stabilization. *J Environ Sci, Al- Mansoura Univ. Egypt, 36,* 281-297.

Abdel-Monem, M. O., Al-Zubeiry, A. H. & Al-Ghithi A. A. (2008). Survival of enteric indicators and pathogenic bacteria in sewage sludge after thermal treatment. *J Bot, Assut Univ. Egypt, 27 (1),* 171-183.

AbdulKarim, M. I., Daud, N. A. & Alam, M. D. Z. (2011). Treatment of palm oil mill effluent using microorganisms. In: M. D. Z, Alam, A. T, Jameel and A, Amid, (eds). Current research and development in biotechnology engineering at International Islamic University Malaysia (IIUM) Vol. III. IIUM Press, Kuala Lumpur, p. 269-275. ISBN 9789674181444.

Adel, A. S., Lalung, J., Efaq, A. N. & Ismail, N. (2015). Removal of cephalexin antibiotic and heavy metals from pharmaceutical effluents using *Bacillus subtilis* strain. *Expert Opin Environ Biol, 4*, 2.

Alam, M. Z., Rashid, S. S., Karim, M. I. A. & Salleh, M. H. (2009). Management of palm oil mill effluent through production of cellulases by filamentous fungi. *World J Microb Biotechnol* DOI 10.1007/s11274-009-0129-9.

Alcalde, M., Ferrer, M., Plou, F. J. & Ballesteros, A. (2006). Environmental biocatalysis: from remediation with enzymes to novel green processes. *Trends Biotechnol, 24 (6),* 281-287.

Al-Gheethi A. A., Norli, I., Efaq, A. N., Bala, J. D. & Al-Amery R. M. (2015) Solar Disinfection and lime treatment processes for reduction of pathogenic bacteria in sewage treated effluents and biosolids before reuse for agriculture in Yemen. *Water Reuse Des* 5(3): 419-429.

Al-Gheethi, A. A. (2015). Recycling of sewage sludge as production medium for cellulase enzyme by a *Bacillus megaterium* strain. *Int. J Rec Org Waste Agri, 4 (2),* 105-119.

Al-Gheethi, A. A., Mohamed, R. M. S., Efaq, A. N., Norli, I., Amir, H. & Ab. Kadir, M. O. (2016). Bioaugmentation Process of Sewage Effluents for the Reduction of Pathogens, Heavy Metals and Antibiotics. *J Water and Health* (Accepted).

Al-Gheethi, A. A. S & Abdul-Monem, M. O. (2014). Effect of nitrogen source on production of cmcase by *Bacillus megaterium* 1295s isolated from sewage treatment plants. *World Acad Sci Eng., Technol Environ Ecol Eng* 1 (6).

Al-Gheethi, AA, Efaq AN, Radin Maya SRM, Norli Ismail and M. O. Ab. Kadir. Potential of Pre-treated Biomass of Bacillus subtilis for Heavy Metal Ions Removal from Aqueous Solution. CLEAN – Soil, Air, Water (Revised submitted).

Al-Gheethi, A. A., Lalung, J., Efaq, A. N., Bala, J. D., & Norli, I. (2015b). Removal of heavy metals and β-lactam antibiotics from sewage treated effluent by bacteria. *Clean Technol. Environ Policy, 17 (8),* 2101-2123.

Al-Gheethi, A. A. S., & Norli, I. (2014). Biodegradation of pharmaceutical wastes in treated sewage effluents by *Bacillus subtilis* 1556WTNC. *Environ Process 1,* 459-489.

Allcock, E. R. & Woods, D. R. (1981). Carboxymethyl cellulase and cellobiase production by *Clostridium acetobutylicum* in an industrial fermentation medium. *Appl Environ Microbiol 41 (2),* 539-541.

Angenent, L. T., Karim, K., Al-Dahhan, M. H., Wrenn, B. A. & Domíguez-Espinosa, R. (2004). Production of bioenergy and biochemicals from industrial and agricultural wastewater. *Trends Biotechnol 22,* 477-485.

Bakare, M. K., Adewale, I. O., Ajayi, A. O. & Shonukan, O. O. (2005). Purification and characterization of cellulase from the wild-type and two improved mutants of *P. fluorescens*. *Afr J Biotechnol 4 (9),* 898-904.

Bala, J. D. (2016). Bioremediation and biodegradation potential of microorganisms isolated from raw palm oil mill effluent Ph.D Thesis, Environmental technology Division, School of Industrial Technology, Universiyi Sains Malaysia (USM).

Bala, J. D., Lalung, J. & Norli, I. (2015). Studies on the reduction of organic load from palm oil mill effluent (POME) by bacterial strains. *Int J Recy Org Waste Agr, 4(1),* 1-10.

Blouzard, J. C., Bourgeois, C., Philip, P., Valette, O., Bélaïch, A., Tardif, C., Bélaïch, J. P. & Pagès, S. (2007). Enzyme Diversity of the Cellulolytic

System Produced by *Clostridium cellulolyticum* Explored by Two-Dimensional Analysis: Identification of Seven Genes Encoding New Dockerin-Containing Proteins. *J Bacteriol 189(6),* 2300–2309.

Boekema, B. K. H. L., Beselin, A., Breuer, M., Hauer, B., Koster, M., Rosenau, F., Jaeger, K. E. & Tommassen, J. (2007) Exadecane and tween 80 stimulate lipase production in *Burkholderia glumae* by different mechanisms. *Appl Environ Microb. 73*, 3838–3844.

Chan, K. Y. & Au, K. S. (1987). Studies on cellulase production by *Bacillus subtilis. Antonie Van Leeuwenhoek 53 (2),* 125-136.

Chynoweth, D. P. & Pratap, P. (1996). Anaerobic digestion of municipal solid wastes: In Palmisano, A. C. and Barlaz, M. A. Microbiology of solid waste. CRC press. Inc. Florida. USA. pp. 71-104.

Davies, J. & Davies, D. (2010) Origins and evolution of antibiotic resistance. *Microbiol. Mol. Biol. Rev. 74 (3),* 417–433.

Davis M. L. & Cornwell D. A. (2013) Introduction to environmental engineering. 15th edition, McGRAW, HILL, International Edition. New York, USA.

Efaq A. N., Nik Norulaini Nik Ab Rahman, Nagao H., Al-Gheethi AA, Md Shahadat & Ab. Kadir M. O. (2015). Supercritical carbon dioxide as non-thermal alternative technology for safe handling of clinical wastes. *J Environ Process 2 (4),* 797–822.

Efaq A. N., Nik Norulaini Nik Ab. Rahman, Nagao, H., Al-Gheethi A.A. & Ab. Kadir M. O. (2016a) Inactivation of *Aspergillus* Spores in Clinical Wastes by Supercritical Carbon Dioxide. *Arab J Sci Eng* (AJSE). Online.

Efaq, A. N., Nik, N. Nik Abd. Rahman, Nagao, H., Alkarkhi, A. M., Al-Gheethi, A. A., Tengku, N. T. L. & Ab. Kadir M. O. (2016) Treatment of solid clinical wastes using supercritical fluid CO_2 as an advanced green technique for destruction of pathogenic fungal spores. *CLEAN – Soil, Air, Water* (Revised submitted).

Ekperigin, M. M. (2007). Preliminary studies of cellulase production by *Acinetobacter anitratus* and *Branhamella* sp. *Afr. J. Biotechnol. 6 (1),* 28-33.

Emtiazi, G., Pooyan, M. & Shamalnasab, M. (2007). Cellulase activities in nitrogen fixing *Paenibacillus* sp. isolated from soil in N-free media. *World J Agri Sci 3 (5),* 602-608.

Fagan, M. J. & Saier, M. H. J. (1994). P-type ATPases of eukaryotes and bacteria: sequence comparisons and construction of phylogenetic trees. *J Mol Evol 38 (1),* 57-99.

Fath, M. J. & Kolter, R, (1993). ABC-transporters: Bacterial exporters. *Microbiol Rev 57(4),* 995-1017.

Gerardi, M. H. (2003). The microbiology of anaerobic digesters. John Wiley and Sons Inc., New Jersey, pp. 91-118.

Gilsdorf, J. R. & Zilinskas, R. A. (2005). New Considerations in Infectious Disease Outbreaks: The Threat of Genetically Modified Microbes. *Clin. Infect. Dis., 40 (8),* 1160-1165.

Hakamada, Y., Koike, K., Yoshimatsu, T., Mori, H., Kobayashi, T., Ito, S. (1997). Thermostable alkaline cellulase from an alkaliphilic isolate *Bacillus* sp. KSM-S237. *Extremophiles 1(3),* 151-156.

Hamer, G. (2003). Solid waste treatment and disposal: Effect on public health and environmental safety. *Biotechnol Adv 22,* 71-79.

Hankin, L. & Anagnostakis, S. L. (1975). Use of solid media for detection of enzymes production by fungi. *Mycologia 67,* 597–607

Herrero, M. & Stuckey, D. C. (2015). Bioaugmentation and its application in wastewater treatment: A review. *Chemosphere 140*, 119–128.

Ikram-ul-Haq, Javed, M. M. & Khan, T. S. (2006). An innovative approach for hyperproduction of cellulolytic and hemicellulolytic enzymes by consortium of *A. niger* MSK-7 and *Trichoderma viride* MSK-10. *Afr J. Biotechnol. 5 (8),* 609-614.

Immanuel, G., Dhanusha, R., Prema, P. & Palavesam, A. (2006). Effect of different growth parameters on endoglucanase enzyme activity by bacteria isolated from coir retting effluents of estuarine environment. *Int J Environ Sci Technol 3 (1)*, 25-34.

Jin, F. & Toda, K. (1989). Nutrients effects on cellulase production by the new species, *Clostridium thermocopriae. Appl. Microbiol. Biotechnol. 31*, 597-600.

Karam, J. & Nicell, J. A. (1997). Potential applications of enzymes in waste treatment. *J. Chem. Tech. Biotechnol. 69 (2),* 141-153.

Kawai, S., Okoshi, H., Ozaki, K., Shikata, S., Ara, K., Ito, S. (1988). Neutrophilic *Bacillus* strain KSM.522 that products an alkaline carboxy methyl cellulase. *Agri Biotechnol Chem 52*, 1425-1431.

Kotchoni, O. S., Shonukan, O. O. & Gachomo, W. E. (2003). *Bacillus pumilus* BpCRI 6, a promising candidate for cellulase production under conditions of catabolite repression. *Afr J. Biotechnol. 2 (6),* 140-146.

Krishna, C. (1999). Production of bacterial cellulases by solid state bioprocessing of banana wastes. *Biores Technol J 69,* 231-239.

Li, J. X., Gao, J., Luo, X. & Guo, Y. (2010). Functions of *Bacillus subtilis* BS7.29 in wastewater treatment. International conference on digital manufacturing and automation (ICDMA) 18-20 Dec. 2010, 1: 753-756.

Lynd, L. R., Weimer, P. J., Van Zyl W. H. & Pretorius, I. S. (2002). Microbial cellulose utilization: Fundamentals and Biotechnology Microbiology. Microbiol *Mol Biol Rev* 66, 506-577.

Miller, G. L. (1959). Use of dinitrosalicylic acid reagent for determination of reducing sugar. *Anal Chem 31,* 426-428.

Mingardon, F., Chanal, A., López-Contreras, A. M., Dray, C., Bayer, E. A. & Fierobe, H. P. (2007). Incorporation of fungal cellulases in bacterial minicellulosomes yields viable, synergistically acting cellulolytic complexes. *Appl Environ Microbiol 73 (12),* 3822-3832.

Morikawa, M. (2006) Beneficial biofilm formation by industrial bacteria *Bacillus subtilis* and related species. *J. Biosci. Bioeng. 101 (1),* 1–8.

Nies, D. H. (1999). Microbial heavy metals resistance. *Appl Microbiol Biotechnol 51 (6),* 730-750.

Prakash, D., Verma, S., Bhatia, R. & Tiwary, B. N. (2011). Risks and precautions of genetically modified organisms. Review Article. ISRN Ecology. 2011: 1-13.

Prescott, V. E., Campbell, P. M., Moore, A., Mattes, J., Rothenberg, M. E., Foster, P. S., Higgins, T. J. V. & Hogan, S. P. (2005). Transgenic expression of bean α-amylase inhibitor in peas results in altered structure and immunogenicity. *J. Agric. Food Chem. 53 (23),* 9023–9030.

Radwan, J. & Babik, W. (2012) The genomics of adaptation. *Proc. R. Soc. B.* p 1-5. (online).

Sadhu, S., Saha, P., Sen, S. K., Mayilraj, S. & Maiti, T. K. (2013). Production, purification and characterization of a novel thermotolerant endoglucanase (CMCase) from strain isolated from cow dung. Springer plus, 2:1-10.

Salmon, D. N. X., Spier, M. R., Soccol, C. R., de Souza Vandenberghe, L. P., Montibeller, V. W., Bier, M. C. J., & Faraco, V. (2014). Analysis of inducers of xylanase and cellulase activities production by *Ganoderma applanatum* LPB MR-56. *Fungal biology*, 118(8), 655-662.

Schloss, P. D., Hay, A. G., Wilson, D. B., Gossett, J. M. & Walker, L. P. (2005). Quantifying bacterial population dynamics in compost using 16S rRNA gene probes. *Appl Microbiol Biotechnol 66,* 457-463.

Sheldon, R. A. & van Rantwijk, F. (2004). Biocatalysis for sustainable organic synthesis. *Austr J Chem 57:* 281-289.

Singh, S., Moholkar, V. S. & Goyal, A. (2013). Isolation, Identification, and characterization of a Cellulolytic *Bacillus amyloliquefaciens* Strain SS35 from Rhinoceros Dung. *Int Scholarly Res Net(ISRN)Microbiol 2013,* 1-7.

Sivaprakasam, S., Mahadevan, S., Sekar, S. & Rajakumar, S. (2008). Biological treatment of tannery wastewater by using salt-tolerant bacterial strains. *Microbial Cell Fact.* 7 (15), 1-7.

Sternberg, D & Mandels, G. R. (1979). Induction of cellulolytic enzymes in *T. reesei* by sophorose. *J Bacteriol 139,* 761-769.

Tao, T., Snavely, M. D., Farr, S. G., Maguire, M. E. (1995). Magnesium transport in *Salmonella typhimurium*: mgtA encodes a P-type ATPase and is regulated by Mg^{2+} in a manner similar of the mgt B P-type ATPase. *J Bacteriol 177 (10),* 2654-2662.

Vinuselvi P., Kim, M. N., Lee, S. K. & Ghim, C. (2012) Rewiring carbon catabolite repression for microbial cell factory. *BMB reports*, 59-70.

Wang, Y. S., Byrd, C. S., Barlaz, M. A. (1994). Anaerobic biodegradability of cellulose and hemicellulose in excavated refuse samples. *J Indus Microbiol 13,* 147-153.

Wang, J. & Chen, C. (2009). Biosorbents for heavy metals removal and their future, Research review paper. *Biotechnol. Adv. 27 (2),* 195–226.

Wu, T. Y., Mohammad, A. W., Jahim, J. M., Anuar, N. (2010). Pollution control technologies for the treatment of palm oil mill effluent (POME) through end-of-pipe processes. *J Environ Manage 91,* 1467–1490.

Yoon, J. H., Park, J. E., Suh, D. Y., Hong, S. B., Ko, J. S. & Kim SH (2007). Comparison of dyes for easy detection of extracellular cellulases in Fungi. *Mycobiol 35(1),* 21-24.

Zhang, Y. H. & Lynd, L. R. (2004). Toward an aggregated understanding of enzymatic hydrolysis of cellulose. Non complexed cellulose systems. *Biotechnol Bioeng 88*: 797-824.

Zhang, Y. H. P., Himmel, M. E. & Mielenz J. R. (2006). Outlook for cellulase improvement: Screening and selection strategies. *Biotechnol Adv 24,* 452-481.

In: Cellulase
Editor: Marguerite Stevenson
ISBN: 978-1-63485-788-8

Chapter 3

CELLULASE PROPERTIES, MARKET AND CHALLENGES OF LIGNOCELLULOSIC MATERIAL HYDROLYSIS

María García-Aparicio[1,2], Michel Brienzo[3,*] and Johann F. Görgens[2]
Department of [1]Microbiology and [2]Process Engineering, Stellenbosch University, Stellenbosch, South Africa
[3]Bioenergy Research Institute (IPBEN), Universidade Estadual Paulista (UNESP), Rio Claro-SP, Brazil

ABSTRACT

The increasing research interest on cellulases have been mainly driven by the production of second generation (2G) biofuels such as ethanol. As such, it is predicted that cellulases experience the largest expansion among the biofuels enzymes. The lignocellulose biomass presents a complex organization making the carbohydrate fraction not easily accessible by cellulases and other auxiliary enzymes. Thus a pretreatment step is necessary to overcome the material recalcitrance, making the carbohydrates more accessible to enzymes action. Progress on pretreatment technologies, reduction of cellulase production cost, and production of more robust cellulases with increased specific activity, have

* Corresponding author: mbrienzo@ipben.unesp.br.

led to considerable reduction of enzyme requirements for lignocellulose hydrolysis. Moreover, the enzymatic hydrolysis optimization based on feedstock and type and conditions of pretreatment applied, can further reduce the enzyme requirements for carbohydrate saccharification into fermentable sugars. Despite these advances, enzyme production and its use are still considered one of the main bottlenecks for commercialization of cellulosic ethanol. A better understanding of the factors influencing the enzymatic hydrolysis of lignocellulose is essential to improve the process. These factors can be grouped into substrate-related and enzyme-related. Enzyme-related factors include the source of enzyme, expression system, biochemical properties of the enzymes, degree of synergism among the cellulolytic components and accessory enzymes, as well as the reaction conditions. In this chapter we will give an overview on cellulase microorganism producers, cellulolytic components and mechanism of action, with a focus on fungal enzymes of *Trichoderma reesei,* since they are predominant on the commercial cocktails used for 2G ethanol production. Factors limiting the enzymatic hydrolysis of lignocellulosic materials and current research trends to counteract them will also be presented.

INTRODUCTION

Biotechnology has increased its weight on industry, which has impelled a slow but growing move toward renewable feedstocks and biological processes. This transition is propelled by the cumulative environmental and economic pressures including climate change mitigation, energy and food security, and resource efficiency. Sustainable practices demand the search of sustainable raw materials (i.e., lignocellulose) and application of novel bioremediation or waste usage processes that minimize environmental impact (i.e., enzyme-based processes). Some biotechnology applications have proven (1) to reduce up to 10-50% in capital and operating cost, (2) to decrease the environmental footprint through savings of up to 10-80% in energy and water as well as the reduction or complete elimination of petrochemical solvents; and (3) to develop new processes for production of new products (OECD, 2001). Precisely enzyme production and its application have given rise to a business that moves billion dollars. Thus their production and utilization has increased steadily in our society contributing considerably to the expansion of biotechnology (Dewan, 2014). Enzymes can be used as final product, as an additive or as industrial biocatalysts. Moreover, new applications are being developed constantly.

Table 1. Uses and benefits of cellulases applications in various industries (adapted from Kuhad et al., 2011)

	Industrial sector	Application	Benefits
	Agriculture	• Plant pathogen and disease control; • Generation of plant and fungal protoplasts;	• Enhanced seed germination and improved root system; • Enhanced plant growth and flowering; • Improved soil quality and reduced dependence on mineral fertilizer
	Feed	• Feed supplementation	• Improved nutritional quality of animal feed, improved feed digestion and absorption.
	Food & Beverages	• Wine production: pressing and color extraction of grapes; • Malting and mashing; beer fermentation, • Maceration, pressing, and color extraction of fruits and vegetables; • Clarification of fruit juices; controlled bitterness of citrus fruits; • Bakery	• Improved aroma of wines; must clarification in wine production; improved filtration rate and wine stability • Improved primary fermentation and quality of beer; improved viscosity and filterability of wort; • Improved viscosity fruit purees; improved texture, flavor, aroma, and volatile properties of fruits and vegetables; • Improved texture and quality of bakery products;
Technical enzymes	Detergents	• Cellulase-based detergents	• Superior cleaning action, improved color brightness and dirt removal; remove of rough protuberances in cotton fabrics; anti-redeposition of ink particle
	Pulp and paper	• Co-additive in pulp bleaching; • Biomechanical pulping; • Enzymatic deinking; • Production of biodegradable cardboard, paper towels, and sanitary paper	• Improved draining; • Reduced energy requirement; • Reduced chlorine requirement; • Improved fiber brightness, strength properties, and pulp freeness and cleanliness; • Improved drainage in paper mills;
	Bioconversion	• Conversion of cellulosic materials to ethanol, other solvents, organic acids and single cell protein, and lipids.	• Reduction of effluents • Reduced formation of inhibitors • Mild conditions
	Textile	• Biostoning of jeans; • Biopolishing of textile fibers;	• Improved fabrics quality; • Improved absorbance property • of fibers; • Softening of garments; • Improved stability of cellulosic fabrics; • Removal of excess dye from fabrics; • Restoration of colour brightness.

Enzymes such as cellulases have been applied as tools for research and development and in almost every industrial sector including agriculture, food, beverages and animal feed, pulp and paper, textile (Table 1). Cellulases is a general term that encompasses multiple enzymes that work in synergism for cellulose degradation. Cellulases belong to the hydrolases type of enzymes (International Union of Biochemistry and Molecular Biology, IUBMB (http://www.chem.qmul.ac.uk/iubmb/enzyme/) able to hydrolyse cellulose molecules by cleaving the β-1,4glycosidic linkages. Cellulases are mainly applied in the food and beverages industries, followed by fine chemicals and its use in detergents (technical enzymes). Cellulases are commonly applied on juice processing together with other glycosil hydrolases such as pectinases and hemicellulases. They are applied during or before the pressing step to loosen the cell walls thereby increasing the juice yield. Cellulases have been used as ingredients of detergents since the 1990s. They are used to modify the structure of cellulose fiber on cotton materials, since they assist in retaining smoothness and color of the fabric. Cellulases are also applied in the textile industry for the production and finishing of cellulose fabrics (biostoning; stonewash denims), biofinishing of cotton and other materials rich in cellulose fibers. These enzymes are also applied to create new types of materials and for effluents treatment. Cellulases also find application in pulp and paper processes such as deinking of recycled waste paper and fiber modification and drain. Moreover, these are the main enzymes applied for production of bioethanol from lignocellulosic biomass (second generation ethanol, 2G ethanol).

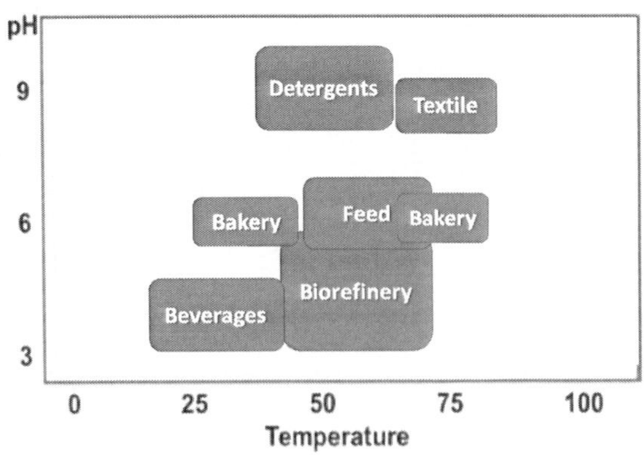

Figure 1. Enzyme conditions of pH and temperature for different cellulase applications.

The different requirements on each application demand specific properties of the cellulases (Figure 1). For instance, the detergent sector looks for cellulases with wide pH range, thermostability and osmotolerance. Effluents treatment and textile industry require cellulases both thermostable and resistant to alkaline conditions. Alkaline cellulases are formed by *Bacillus* strains, and neutral and acidic cellulases are produced by *Trichoderma* and *Humicola* fungi. In the case of the biorefinery industry, thermostability and resistance to acidic conditions are attributes desired in the cellulases.

Lignocellulose biomass constitutes a renewable and abundant feedstock for the production of biofuels, biomaterials and bioproducts. Lignocellulose can be obtained at relatively low cost from different resources and industries (Cherubuni, 2010): agriculture (dedicated crops and residues), forestry and pulp and paper industries (process residues and leftovers) and aquaculture (algae and seaweeds). These materials are composed basically of three structural components namely cellulose, hemicellulose and lignin, that can be obtained through several fractionation processes. The polysaccharides, cellulose and hemicellulose, can be hydrolyzed into monomeric sugars that can be fermented into biofuels such as ethanol. The lignin fraction is mostly used as source of energy (electricity, steam) for the lignocellulose conversion process, although lignin and its derivatives can be used to produce adhesives, coatings, paints, and plastics of renewable origin (Imman et al., 2013). Lignocellulose conversion technologies for biofuels (bioethanol) production can be based on acid hydrolysis or enzymes (biocatalysis). Concentrated acid hydrolysis process is carried out at moderate temperatures and it provides high sugar yields. However, the process is expensive owing to the large amount of acid required (30-40%), high energy consumption for catalyst recovery and requirement of special equipment highly resistant to corrosion. The diluted acid version requires less acid but higher temperatures, what promotes corrosion in the equipment and higher degree of degradation of hemicelluloses derived sugars (Taherzahed & Karimi, 2007). Dilute acid hydrolysis, however, is applied as pretreatment prior to enzymatic hydrolysis (Alvira et al., 2010a). Although concentrated acid hydrolysis is being considered industrially (Arkenol, Masada resource Group, Ethtec at pilot scale), the hydrolysis of lignocellulose via enzymes is considered a more attractive alternative from a technological and environmental point of view. This fact is highlighted by the larger number of second generation plants (pilot, demonstration and commercial plants) based on enzymatic saccharification (Table 2). Enzymatic catalysis offer several advantages compared to their chemical counterparts. They require milder conditions of pH and temperature, are substrate specific and do not result in formation of inhibitory compounds that require neutralization prior to fermentation.

Table 2. Pilot, demonstration and commercial plants for second generation (2G) ethanol production (database of the Biofuels Digest, http://www.biofuelsdigest.com/)

Company/Project	Commercial Plants	Location	Feedstock	Saccharification Technology
Abengoa	3	USA, Brazil	Corn stover, bagasse	Enzymatic hydrolysis
Beta-Renewables	2	USA, Italy	Mixed biomass	Enzymatic hydrolysis
BlueSugars	1	Brazil	Stover	Enzymatic hydrolysis
BlueFire Renewables	1	USA	Municipal Solid Wastes	Acid Hydrolysis
BP Biofuels	2	USA	Bagasse, miscanthus	Enzymatic hydrolysis
Butalco	Pilot	Germany	Lignocellulose	-
China Resources Alcohol Corporation (CRAC)	Pilot	China	Corn stover (stalks and leaves)	Enzymatic Hydrolysis
COFCO/Sinopec	1	China	Corn stover	Enzymatic hydrolysis
Dupont Biofuels Solutions	2	USA	Corn cob	Enzymatic hydrolysis
Ethtec	Pilot	Australia	Wood residues	Acid Hydrolysis
GraalBio Investimentos	6	Brazil	Bagasse	Enzymatic Hydrolysis
Inbicon	2	Denmark	Straw	Enzymatic Hydrolysis
Iogen Corp	Demo	Canada	Wheat Straw	Enzymatic Hydrolysis
Lignol	Pilot	Canada	Wood wastes	Enzymatic Hydrolysis
Mascoma	1	USA	Hardwood	Consolidated Bioprocessing
POET-DSM	9	USA	Corn stover, cobs	Enzymatic Hydrolysis
MO Renewables - COFCO/CNOOC	1	China	Cellulosic Ethanol	Enzymatic Hydrolysis
Weyland/Statoil Hydro	1	Norway	Cellulosic Ethanol	Enzymatic Hydrolysis

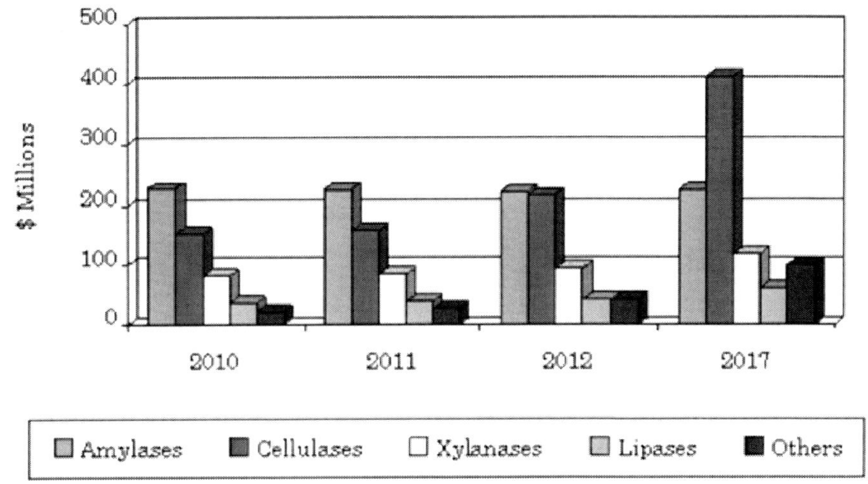

a

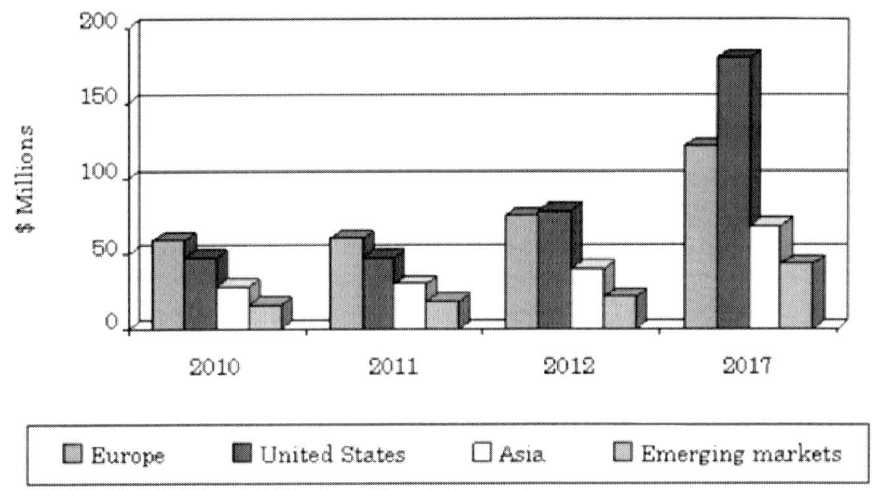

b

Figure 2. (A) Global market for biofuel enzymes and (B) sales for cellulases in the biofuels sector for different regions for the period 2010-2017 ($ millions) (from Dewan, 2013).

The increasing research interest on cellulases have precisely been driven by the production of second generation (2G) biofuels such as ethanol. In enzyme-based process, cellulases and other auxiliary enzymes are used for the saccharification of cellulose and other polysaccharides in plant biomass to monomeric sugars for biofuel production. However, 2G bioethanol production requires a complex processing technology compared to that applied on sugar-rich or grain biomass for 1G bio-ethanol production (Sánchez & Cardona, 2008). Although this fact has hampered the commercial production of 2G bio-ethanol, the continuous progress in research and development is underlined by the strong investments in production capacity to meet fuel replacements targets, government support of several countries (USA, Brazil) with funding of research and development in the sector (Table 2). Thus, the demand for bioethanol enzymes, mainly cellulases, has experienced a strong growth. The market for cellulases is predicted to grow at a compound annual growth rate (CAGR) of 13.2% with profits estimates of up to $412 million for 2017 (Figure 2) (Dewan, 2013). Although Europe presented the largest share of 38.5% in the cellulase market in 2011, United States is predicted to experience the greatest profits with an estimated 43.6% of the global cellulase sales (Figure 2).

Table 3. Leading producers and distributors of cellulases (from Dewan, 2013)

Company	Product
Codexis Inc	CodeXyme cellulase
DuPont/ Danisco/ Genencor	Accellerase TRIO Accellerase 1000 Accellerase 1500 Accellerase DUET Accellerase XC Accellerase BG
Dyadic	AlternaFuel CMAX AlternaFuel 100P AlternaFuel 200P
Iogen Coorporation	Ultimase XTP
Montana Microbial products	MMP cellulase
Noor Creations	BioConvert –ACEL (L)
Novozymes	CellicCTec2 CellicHTec2 CellicCTec3 CellicHTec3
SEB	SacchariSEB C6

Table 4. Cellulolytic system of *Trichoderma reesei* (from Seiboth et al., 2011)

GH Family	CAZy nomenclature	Previous designation	Cellulase type	Size (aminoacids)	Position of CBM	Stereo-selectivity
1	Cel1A	BGL1	β-glucosidase	466	-	Retaining
1	Cel1B		β-glucosidase	484	-	Retaining
1	Cel3A		β-glucosidase	744	-	Retaining
3	Cel3B		β-glucosidase	874	-	Retaining
3	Cel3C		β-glucosidase	833	-	Retaining
3	Cel3D		β-glucosidase	700	-	Retaining
3	Cel3E		β-glucosidase	765	-	Retaining
5	Cel5A	EG2	Endoglucanase	397	N	Retaining
5	Cel5B			438	GPI anchor	Retaining
6	Cel6A	CBH2	Cellobiohydrolase	447	N	Inverting
7	Cel7A	CBH1	Cellobiohydrolase	497	C	Retaining
7	Cel7B	EG1	Endoglucanase	436	C	Retaining
12	Cel12A	EG3	Endoglucanase	218	-	Retaining
45	Cel45A	EG5	Endoglucanase	270	C	Inverting
74	Cel74A	EG6	Endoglucanase/ xyloglucanase	818	C	Inverting

Diversa, Novozymes, DuPont (after gaining Danisco-Genencor) and Dyadic International produce enzymes for second generation ethanol. AB Enzymes, Advanced Enzyme Technologies, DSM, Montana Microbials and Noor Creations are other companies that produce cellulases for 2G biofuels manufacturing. The enzyme market is currently controlled by Novozymes and DuPont, being Novozymes the leader with about 53% of the total share on 2011. The leading producers and supplier of cellulases for 2G biofuels production are listed in Table 3.

The production of ethanol from lignocellulose generally consists of a pretreatment, enzyme production, enzymatic hydrolysis, fermentation and distillation. Efficiency of cell wall saccharification is influenced by many factors encompassing feedstock characteristics (hardwood, softwood, herbaceous), type and conditions of pretreatment and enzymatic hydrolysis, including the solids loading, source of enzymes and combination employed thereof (Pengilly et al., 2015). In this chapter we will give an overview on cellulase microorganism producers, cellulolytic components and mechanism of action with focus on fungal enzymes of *Trichoderma reesei,* since they are predominant on the commercial cocktails used for 2G ethanol production. Factors limiting the enzymatic hydrolysis of lignocellulosic materials and current research trends to counteract them will also be presented.

SOURCE OF CELLULASES ENZYMES

In nature, several organisms can degrade lignocellulosic materials through producing complementary enzymes that work on synergism. The lignocellulose breakdown is reached under very different environmental conditions in terms of pH, redox, temperature and pressure. As such, lignocellulose degrading enzymes such as cellulases are produced by a great variety of organisms including superior plants, some invertebrates and mainly micro-organisms (Figure 3). Superior plants produce cellulases to weaken cell wall so as to facilitate growth, differentiation or abscission at different stages of the plant development. Cellulases are also produced by pathogenic organisms allowing the penetration and invasion of plant tissues. In the case of microorganisms, enzymes such as cellulases are required to use cellulose as carbon source. Examples of these micro-organisms are aerobic actinomycetes, aerobic filamentous fungi, anaerobic fungi and anaerobic hyperthermophilic bacteria (Cragg et al., 2015). Some bacteria genera are *Acetovibrio*, *Actinomycetes*, *Bacillus*, *Bacteriodes*, *Cellulomonas*, *Clostridium*, *Erwinia*, *Microbispora*, *Pseudomonas*, *Ruminococcus*, *Streptomyces*, *Thermobifida* and *Thermomonospora*. The major cellulases among fungi are aerobic from different genus such as *Aspergillus*, *Fusarium*, *Humicola*, *Monilia*, *Penicillium*, *Sclerotium*, *Schizophyllum*, *Talaromyces* and *Trichoderma*.

Cellulases from extremophiles have great potential owing to their ability to catalyse reactions under extreme environmental conditions of pH, temperature or salinity. One of the promising areas for novel enzyme discoveries for industrial applications is the marine environment (Lawson, 2011). The ocean conditions of salinity, pressure, temperature and light make this environment a potential source of enzymes that are more stable that the terrestrial equivalents. Other extremozymes have been isolated from microorganisms found in Arctic, deep-sea volcanoes and high salinity environments. Thermostable cellulases have been obtained from bacteria isolated from compost systems, soils, wastewaters and deep subsurface of goldmines (Kuhad et al., 2016). These cellulolytic systems present interesting properties from an industrial point of view. Most industrial processes, e.g., second generation (2G) ethanol production, would benefit of thermostable enzymes that generally have an increased stability and specificity, which in turn would result in reduced enzyme dosage required for the process (Viikari et al., 2007). Performing processes at high temperatures presents other advantages such as higher conversion rates, reduction of risk of contamination, reduction of viscosity, increased solubility of polymers and improved bioavailability (Gomes & Steiner, 2004). On the other hand, the use of enzymes able to work efficiently at milder temperatures could allow a better flexibility of process configurations matching optimum temperatures of enzymatic hydrolysis and fermentation processes (Wang et al., 2013). Some bacteria from the genera *Clostridium* synthetize an extracellular cellulolytic system, with the ability of degrade both cellulose and hemicellulose (Bayer et al., 2004; Yammamoto & Tamaru, 2016). Novel multifunctional enzymes (exo and endoglucanase, xylanase activities) with resistance to ionic liquids, salt and inhibitors have been identified in cow rumen microbial community (Loaces et al., 2016).

However, the majority of enzymes that have been used in biotechnology derived mainly from filamentous fungi, *Aspergillus*, *Fusarium*, *Phanerochaete* and *Trichoderma*, owing to their high secretion titres, facilitating the further enzyme recovery and purification from the fermentation media (Persson et al., 1991; Conesa et al., 2001; Banerjee at al., 2010). Moreover, fungal enzymes are generally less complex compared to those from bacteria, which facilitates the cloning and production through recombination in a fast grower bacterial host such as *Escherichia coli* (Kuhad et al., 2016). Precisely, the most studied and exploited microorganism for industrial cellulase production has been the actinomycete *Trichoderma reesei* (anamorphous form of *Hypocrea jecorina*), that was discovered during the Second World War because it was responsible of deterioration of cotton fabrics of the US soldiers. *T. reesei* has undergone multiple rounds of strain improvement since 1950 (Banerjee at al., 2010; Olofsson et al., 2008; Samuels, 2006), and it is currently the main industrial cellulolytic enzyme producer.

NOMENCLATURE AND DATABASES

Cellulase enzymes belong to the glycoside hydrolase (GH – 135 families,) that share their ability to hydrolyze glycosidic bonds between two or more carbohydrates, or between a carbohydrate and a non-carbohydrate moiety. The numeric code EC (enzyme catalogue) assigned to the GH by the "Internacional Union of Biochemistry and Molecular Biology" (IUPAC) is EC 3.2.1. – where the first digit indicates the class of enzyme (hydrolase), the second digit refers to the subclass (capacity of hydrolyse glycosidic bonds), the third digit indicates the donor group (water) and the fourth number corresponds to the acceptor molecule that is substrate specific. In the case of fungal enzymes, the cellulose system consists of mainly three enzyme activities namely exoglucanase (cellobiohydrolase, ß-(1,4)-glucanglucohydrolase, EC 3.2.1.74 and EC 3.2.1.91), endoglucanase (1,4- ß-glucanglucanohydrolase (EC 3.2.1.4) and glucohydrolase (β-glucosidase, EC 3.2.1.21) (Himmel et al., 1996).

Rigorously speaking, EC numbers do not correspond to a specific enzyme but to enzyme-catalyzed reactions, meaning that different enzymes that conduct the same reaction will share the same EC number. Although the EC classification has been used broadly, it presents some limitations since it does not consider the specific enzyme structure, its amino acid sequence or mechanisms of action, all of which are significant parameters for elucidating the enzymatic reactions. Thus, recent enzyme databases have been created to cover a wide range of aspects such as enzyme function and properties, structure (three-dimensional folds) and amino acid sequence (Schomburg& Schomburg, 2010). Carbohydrate Active enZYme (CAZy) database has specialized in carbohydrate acting enzymes such as glycosil hydrolases and other accessory enzymes for carbohydrate depolymerization. This resource has been very useful to elucidate novel glycosil hydrolases through bioinformatics analysis. This system is used to establish phylogenetic relationships among the different families. This type of classification also allows for prediction of catalytic mechanisms and catalytic motifs since they are conserved in most of the GH families. Although the enzymes from the same group share similar 3D structure, mechanism of reaction and catalytic residues, they differ on substrate specificity. Some members of this family have been grouped into clans (from GH-A to GH-N) according to structural folding (and catalytic mechanism), that is more conserved than the actual amino acid sequence. The CAZy database is updated regularly, it has catalogued 135 families up to date (http://www.cazy.org/Glycoside-Hydrolases.html), while other families like GH-61 has been reclassified as auxiliary enzymes AA9 (Corrêa et al., 2016; Levasseur et al., 2013).

Cellulolytic System Composition and Molecular Structure

The filamentous fungus *Trichoderma reesei* is nowadays a model for commercial scale production of cellulases. This fungus can produce more than 100 g/L of secreted protein under industrial fermentation conditions (Druzhinina & Kubicek, 2016). *T. reesei* secretes different types of cellulolytic enzymes, although the main components are exoglucanases. Two of these enzymes have been identified in *T. reesei*, each occurring in several iso-enzymatic forms as CBH I (Cel7A, about 60%) and CBH II (Cel6A, 20%). The most common cellulases of *T. reesei* with the corresponding CAZy nomenclature and specific stereospecificity are listed in Table 4. These enzymes work together in synergism to hydrolyse cellulose by creating new available regions for each other, removing any obstacles and alleviating end-product inhibition (glucose, cellobiose) that may occur. The endo-glucanases (EG) cleave randomly internal β-1,4-glucosidic bonds in amorphous regions of the cellulose, reducing significantly the degree of polymerization of cellulose (Jørgensen et al., 2007). Exoglucanases (CBH) bind to both reducing and non-reducing ends of the cellulose chains, processively cleaving off cellobiose units without desorption of the cellulose after each catalytic event. The activity of β-glucosidases is required to hydrolyse the cellobiose units to glucose. GH families can be grouped into two classes according to their stereoselectivity, reverting or inverting (Table 4), based on the configuration of the anomeric oxygen during the reaction (Vuong &Wilson, 2010). Commonly, the retaining enzyme has typical acid/base residue and a nucleophile, whereas the inverting enzyme contains both catalytic acid and base residues.

Most of cellulases of *T. reesei* (except for EG III or Cel12A) and other fungal cellulases present a similar structure based on two differentiated domains linked by a flexible short peptide of 22-34 aminoacids (Rabinovich et al., 2002). Cellobiohydrolases have a small C-terminal glycopeptide (carbohydrate binding domain or module, CBD or CBM) that is able to bind to cellulose as well as a larger core protein (catalytic domain, CD), which contains the active site (Olsson et al., 2005). The CBM can be positioned at the N- or C-terminal end of the protein (CBH2 and CBH1, respectively). The role of the CBM is to increase the concentration of the enzymes on the substrate thereby increasing contact time by bringing the catalytic core into close contact with the surface of the celluloses (Quentin et al., 2003). The presence of CBM enhances cellulase reactivity when it presents moderate binding affinity (Walker et al., 2015) on the crystalline region of insoluble cellulose substrates (Lemos et al., 2003; Thongekkaew et al., 2013). On the other hand, enzymes without CBM are able to easily desorb from the lignocellulosic biomass substrates, and thus preventing "traffic junks" in the cellulose chain. However, recent studies on enzymatic hydrolysis at high solids content have demonstrated that the enzymes act as efficiently with and without CBM (Várnai et al., 2014). The CBMs have been also grouped according to their amino-acid sequence into 74 families in the CAZy database (http://afmb.cnrs-mrs.fr/CAZY/fam/acc_CBM.html, May 2016). The different structure of the CD of cellobiohydrolases and endoglucanases is responsible of the specific mechanism of action. The CD of the cellobiohydrolases consists of a tunnel-shaped structure of 20-50 Å length (CBHII-CBHI) in which the active site is positioned (Figure 4). In the case of the endoglucanases, the catalytic domain is cleft-shaped, thus exposing their active sites to the external part of the enzyme (Henriksson et al., 1996).

Other cellulolytic enzymes of some rumen bacteria (*Acetovibrio*, *Clostridium*, *Bacteroides*, *Ruminicoccus*) and anaerobic fungi (*Neocallimastix*, *Piromyces* and *Orpinomyces*) are arranged in a multimodular structure named cellulosome, where several catalytic domains are organized on a scaffold. Most cellulosomes present a similar structure composed of scaffolding proteins (cohesion and CBM domains), hydrophilic domains, dockerin domains, enzyme subunits and domains with unknown function (Figure 5). Other multimodular cellulases with more than one active unit have been described such a cellulase containing GH5, GH6, CBM5 and CBM domains (Eckborg et al., 2007).

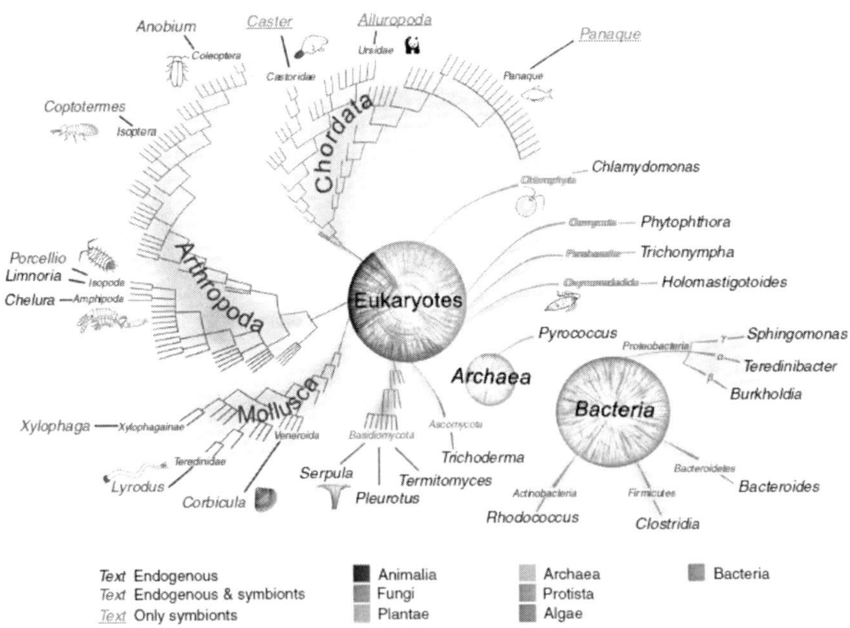

Figure 3. Distribution of organisms with the ability of degrading lignocellulose (from Cragg et al., 2015).

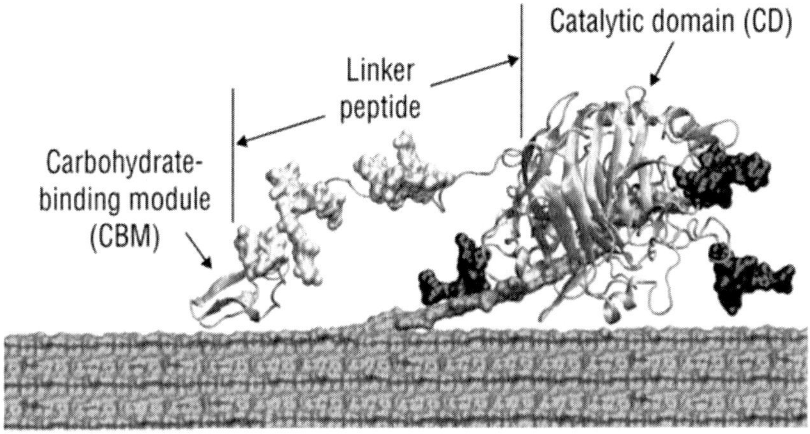

Figure 4. 3D structure of the Cel7A (CBH1) of *T. reesei* (source: NREL).

Mechanism of Action

The simplified mechanism of action that is generally considered for cellulases is the sequential and synergetic interaction of the main three activities, namely CBH, EG and β-glucosidase (Medve et al., 1998). This simplified model, however, does not explain the reduction of synergism observed during the enzymatic hydrolysis process. Recent models have demonstrated the processive action of cellobiohydrolases, by which the enzyme advances through the same cellulose chain, while cleaving cellobiose units and exposing the cellulose chain above (Bekham et al., 2014). Endoglucanases cleave not only cellobiose, but also cellodextrines and glucose, which suggests a pseudo-processive mechanism by which several glycosidic bonds are hydrolysed in the cellulose chain. Moreover, EGs can reach adjacent cellulose chains while bonded thanks to the flexible peptide linker. Thus, the hydrolysis mediated by these enzymes is not as random as originally thought. Besides working on amorphous regions of cellulose, EGs would prevent the confinement of CBHs in shortened cellulose chains (Eriksson et al., 2002). There is also synergism between the CBHs (exo-exo synergism) of *T. reesei*, since the CBHI (Cel7A) acts on the reducing end of the cellulose chain while the CBHII (Cel6A) works on the non-reducing end. EGs and CBHs exhibit synergism towards *β-glucosidase, since it cleaves the cellobioses that inhibit the activity of CBH (*Xiao et al., 2004*;* Yue et al., 2004*).*

*Other enzymes that can boost cellulase activity are hemicellulases (xylanases, mannanases, arabinofuranosidases), esterases (acetyl, ferulic acid and glucuronoyl esterases), "swollenin" and oxidative enzymes [cellobiose dehydrogenase, lytic polysaccharide monooxygenases (LPMO)]. Precisely, one of the strategies to improve carbohydrate conversion yields during enzymatic hydrolysis is the development of tailor-made enzyme combinations including different enzyme activities. It is believed that an optimal combination of enzymes working in synergism could result in a reduction of the total enzyme dosage and/or reduction of hydrolysis time (*Pengilly et al., 2016*). For example, the addition of endo-xylanases has been proven to promote cellulose conversion by cellulases (*García-Aparicio et al., 2007*;* Hu et al., 2011*;* Pengilly et al., 2015*). There are several factors that could explain the positive interaction of cellulose and xylanase including removal of residual xylan and disruption of lignin-carbohydrate linkages that constitute a barrier to cellulases; hydrolysis of xylo-oligosaccharides, potent inhibitors of cellulases; binding to lignin thereby reducing unproductive binding of cellulases; augment porosity and swelling of cellulose fibers, what in turns increase the available surface area for cellulases. Similarly, swollenin, a protein with non-catalytic activity and also secreted by T.reesei, could also act altering cellulose fibers increasing their accessibility (*Saloheimo et al., 2002*).*

*Lytic polysaccharide monooxygenases (LPMOs), classified as auxiliary enzymes AA9 in CAZy database, have been proven to boost cellulose activity by creating new ends for the cellulases. LPMOs (previously GH61) are oxidative enzymes that contain plane substrate-binding surfaces and cleave polysaccharide chains in their crystalline form. Their activation depends on the presence of reductants that act as an electron donor, non-enzymatic or enzymatic, as well as divalent metal ions (*Horn et al., 2012; Martínez et al., 2016*) (Figure 5). Cellobiose dehydrogenase (CDH), another redox enzyme from the auxiliary family AA3, fulfills this function by the oxidation of cellobiose and transfer of electrons to the heme group of the LPMO. Likewise, glucose dehydrogenase (GDH) has been proven to be effective in LPMOs activation by the oxidation of glucose using redox recycling by quinones. Phenolic compounds from the lignin and plant extractives can also act as reductant molecules for LPMO activation, but their recycling through LPMO oxidation results in radicals with low oxidative capacity *(Martínez et al., 2016*). Photosyntetic pigments can also act as non-enzymatic reductants and get back to the reductant state by lignin compounds. The necessity of both sugars and lignin in the activation of LPMOs makes evident the close association between polysaccharide and lignin degradation.*

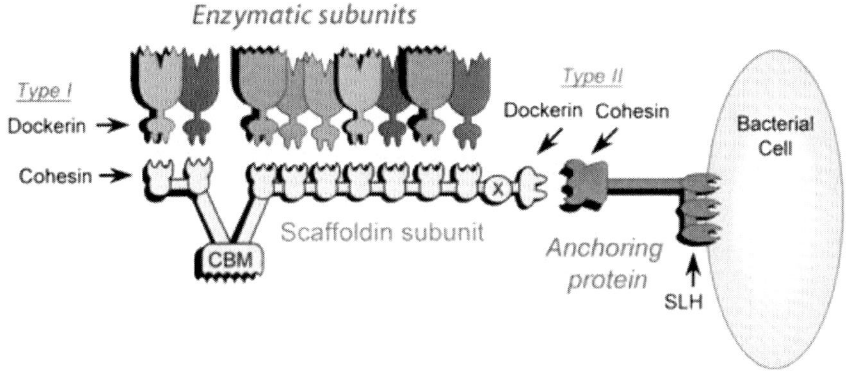

Figure 5. Architecture of the C. thermocellum cellulosome system (https://www.cazypedia.org/index.php/Cellulosome).

CELLULASE PRODUCTION

Cellulase production cost impact considerably in the economics of bioethanol production from lignocelluloses. The composition of growing media and conditions of the enzyme-production bioprocess will determine the enzyme yield. Several factors such as carbon source, the pH, agitation, light, fungi morphology, metabolism induction, to name a few, affect enzyme titres (Domingues et al., 2000; Gyalai-Korpos et al., 2010). Commercial cellulases are generally obtained through submerged fermentation of *T. reesei.* Submerged fermentation requires rich fermentation broth and a high concentration of oxygen. This production system allows for a better control of fermentation parameters, but the highly nutritive media contributes considerably to the cost. An interesting alternative is the use of the pretreatment liquor as carbon source (Lo et col., 2005). Alternatively, enzyme production can be carried out using solid substrates such as wheat bran and other lignocellulosic biomass (Juhász et al., 2005; Jorgensen & Olsson, 2006). This alternative can be easily integrated in 2G ethanol facilities as some raw material can be diverted for enzyme production (Lo et al., 2005). Crucial parameters to be controlled during this fermentation are the particle size and the level of moisture retained in the substrate. Besides the relatively low cost of feedstock, solid state fermentation presents additional advantages including simple fermentation equipment requirements, reduction on effluents generated, lower catabolite repression and higher yields. Moreover, the enzyme cocktail synthetized would be substrate specific (Tengerdy & Szakacs, 2003). Despite the advantages, the solid state fermentation process suffers from the absence of suitable technology for the scale up at industrial level, purification of the end-products and biomass requirements. One of the main problems yet to be solved is the control of heat transfer, although several reactors designs that include cooling systems offer promising results (Kuhad et al., 2016).

FACTORS THAT INFLUENCE ENZYMATIC SACCHARIFICATION

The biochemical production of ethanol from lignocellulose essentially consists of the following steps: pretreatment, enzymatic hydrolysis, fermentation and product recovery. The pretreatment is the step considered as the most critical stage during the biological conversion to ethanol, having a large influence on the enzymatic hydrolysis and fermentation (Alvira et al., 2010a). Although pretreated materials are more amenable to enzymatic digestion, the enzyme production and its use during enzymatic hydrolysis can add up to 25% of the final expenses (Bansal et al., 2012). This is due to the lower specificity of the lignocellulolytic enzymes compared to others biofuel enzymes (e.g., amylases used for first generation ethanol production from grain), making necessary to use larger amounts and longer incubation periods for a complete conversion. Techno-economic analysis indicate that the enzyme dosage required to reach cellulose conversions superior to 90% may not compensate for increased enzyme cost contribution (Arantes et al., 2011). Subsequently, fixing a target where 10–20% of the cellulose is not hydrolysed may be a more economic approach. Moreover, long incubation periods increase significantly operational cost and thus the overall conversion process (Sheng et al., 2008; Newman et al., 2013). Thus, economic analysis is required to determine the tradeoff among yield (cellulose conversion), enzyme dosage and incubation time.

The reaction catalysed by cellulases is initially a solid-liquid system consisting of an insoluble substrate (pretreated lignocellulose) and a soluble catalyst (cellulolytic enzymes). Both the structural characteristics of the lignocellulose and the mechanism of action of the enzymes combination will impact on the rate of the conversion of cellulose. Nevertheless, the cellulose conversion into glucose normally follows three different stages characterized by fast, moderate and low rates of hydrolysis (Wallace et al., 2016). Figure 7 shows a typical progress of the glucose release over time during enzymatic hydrolysis of pretreated lignocellulose with three different enzyme preparations. Cellulases first adsorb quickly into the accessible cellulose, where after a rapid, initial hydrolysis phase takes place. The reaction, however, rapidly reaches an intermediate phase that is characterized by the moderate hydrolysis of the substrate. Between 50 and 70% of the initial cellulose has been hydrolysed at that point, depending on substrate (feedstock, pretreatment applied, concentration) and enzyme (source of enzyme, loading). The last phase corresponds with the steady decrease of the reaction rate, owing to the recalcitrant nature of the remaining polysaccharides (Arantes et al., 2011).

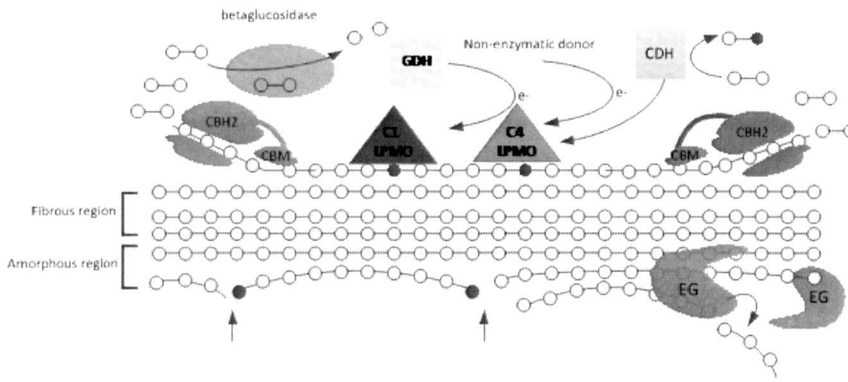

Figure 6. Synergistic fungal enzymatic degradation of cellulose. CBM – carbohydrate-binding module, CBH - cellobiohydrolase, CDH – cellobiose-dehydrogenase, EG – endo-glucanase, GDH-glucose dehydrogenase, LPMO –lytic polysaccharide monooxygenase (Figure adapted from Horn et al., 2010).

There are several other factors involved in the reduction of hydrolysis rate during cellulose conversion. These factors can be divided into substrate- and enzyme-related factors. Depending on the type of biomass, pretreatment and enzymes applied, enzyme- or substrate-related factors could exert bigger influence than the other (Del Rio et al., 2012). The extent of reduction of the enzymatic hydrolysis yield will be determined by the following factors (Figure 8):

- Recalcitrant and heterogeneous nature of substrate: crystallinity of cellulose, degree of polymerization, presence of hemicellulose and acetyl groups (steric hindrance), porosity, etc.

- Properties of the cellulolytic system: nature, size, mechanism of action, structure (cellulose binding (CBD) domain).
- Conditions of the enzymatic hydrolysis (pH, temperature, agitation, incubation period, enzyme dosage, substrate loading) that could lead to deactivation of the enzymes due to temperature and shearing forces.
- Enzyme inhibition:
    - End-product inhibitions by glucose and cellobiose.
    - Inhibition by oligosaccharides generated during hydrothermal pretreatment or released by side activity enzymes.
    - Inhibition or inactivation by compounds generated during pretreatment.
- Unproductive binding of cellulase to lignin or entrapment of cellulases between cellulose chains.
- Loss of synergism between enzyme components.
- Increased recalcitrance of the substrate as the hydrolysis proceeds (since the amorphous regions are hydrolysed, the substrate gets enriched in the crystalline cellulose and lignin-carbohydrate complexes).

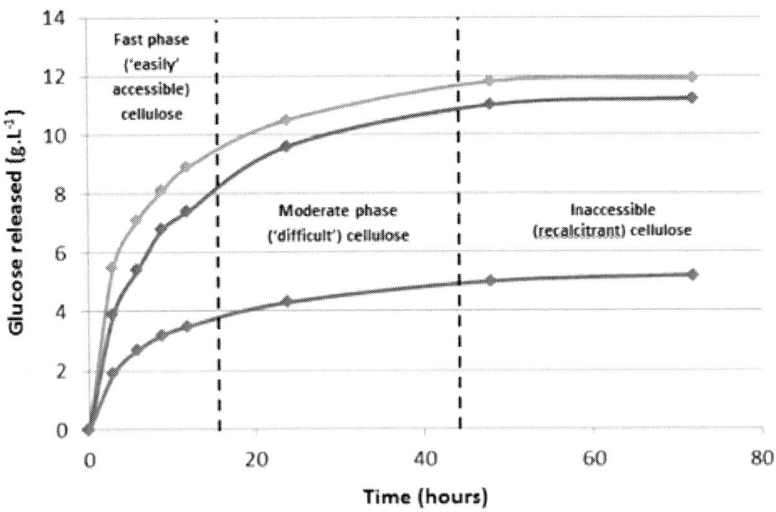

Figure 7. Typical time course for glucose release during enzymatic hydrolysis of pretreated lignocellulose with different commercial cellulases. Substrate: steam exploded triticale straw, 2% WIS (w/v).

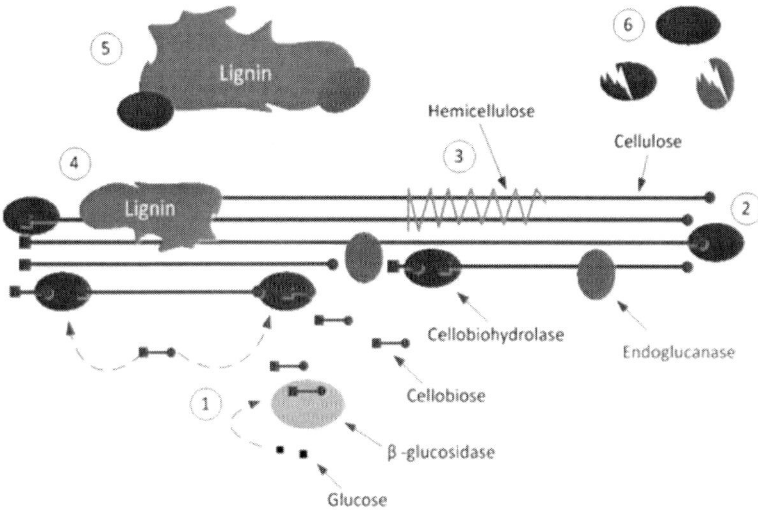

Figure 8. A simplistic overview of the factors that limit the efficient hydrolysis of cellulose. 1: Product inhibition of the cellobiohydrolases and β–glucosidases by cellobiose and glucose, respectively. 2: Unproductive binding of the cellobiohydrolases onto a cellulose chain. Due to the strong binding of the cellulose chain by cellobiohydrolases in their catalytic core, obstacles can make the enzymes pause and thereby become unproductively bound. 3 and 4: Association or covering of the microfibrils by the hemicelluloses and lignin can prevent access of the cellulases to the cellulose surface. 5: Both cellulase and hemicellulase enzymes can adsorb un-specifically onto lignin particles or surfaces. 6: Loss or denaturation of the enzyme activity due to mechanical shear, low thermostability or proteolytic activity (Figure adapted from Jorgensen et al., 2007).

Research in Enzyme Technology for Lignocellulose Conversion

Efficient hydrolysis of polysaccharides of lignocellulosic biomass into their fermentable units (monomers) is fundamental for cost-effective utilization of the lignocellulosic feedstock (Olofson et al., 2008). Their recalcitrance (intrinsic resistance to deconstruction) constitutes the main barrier for the development of biochemical technologies for 2G biofuels production (Meyer et al., 2009). Progress on pretreatment technologies, reduction of cellulase production cost, and production of more robust cellulases with increased specific activity have led to considerable reduction of enzyme requirements for lignocellulose hydrolysis. Genencor (now Dupont) and Novozymes claimed a reduction of 20-30 fold on production cost as well as increased of specific activity of the cellulase cocktails (Zhang, 2011). Despites these advancements, production and application of cellulases need to be improved (Klein-Marcuschamer et al., 2012). Existing enzyme production costs have been estimated between $ 5/Kg protein (Zhang, 2011) and $10/Kg protein (Klein-Marcuschamer et al., 2012), while predictions from Biomass DOE Program suggest a reduction of the enzyme production cost to 0.7$/Kg of protein. Further reductions on cost can be achieved by the improvement of the enzymatic hydrolysis step so that the cellulase requirement are reduced.

As the cost of enzyme production and its use are one of the major contributors in the final cost, substantial research is being directed towards the improvement of enzymatic hydrolysis. Among the different areas of research and aspects that are directly involved in the hydrolysis efficiency should be pointed out the following (Viikari et al., 2012):

- Search of *new feedstocks with desirable properties* for the process such as higher productivity per hectare, low inputs requirement for production, high carbohydrate content, reduced lignin content and reduced recalcitrance (Benjamin et al., 2014).
- Search of *novel pretreatment technologies* and enhancement of existing pretreatments and its efficiency so that there is a reduction in total enzyme requirements for carbohydrate hydrolysis. The partial removal or relocation of lignin during pretreatment minimizes the unproductive binding of cellulases. Recent pretreatments utilize green solvents such as ionic liquids, most of which can be conducted at mild conditions while they can be recovered after the process. Some ionic liquids containing chloride and bromide, however, result in deactivation of cellulases. Current trends of research are the search of novel cellulases with high tolerance to ionic liquids as well as the development of ionic liquids compatible with cellulases. The latter will allow for a combined pretreatment and enzymatic hydrolysis in one pot (Elgharbawy et al., 2016).
- Reduction of enzyme production cost by:
    - *Integrated near-site or in situ enzyme production* systems that use cheaper carbon sources, including lignocellulosic substrates (Johnson, 2016).
    - Improvements in cellulase expression by:
        - Development of enzyme-producing strains with improved protein-titres through heterologous expression in different

expression systems including bacteria (*Escherichia coli*) and yeast (*Saccharomyces cerevisiae* and *Pichia pastoris*). Filamentous fungi have also been used as expression system owing to their intrinsic ability to produce great amounts of proteins. The use of the strong promoters such as the one of CBH-I, has resulted in significant improvements of the secretion of cellulolytic enzymes in *T. reesei*. Other studies demonstrate the improvement of cellulase synergism by the coproduction of several cellulolytic enzymes by using multipromoter and multicistronic vectors (Kuhad et al., 2016).

The screening of improved and/or novel enzymes requires of high through put (HTP) systems to evaluate the cellulase activity. A major challenge during the development of such systems is the difficulty in handling the material at milligram scale, as lignocellulosic biomass is a heterogeneous and insoluble material (Berlin et al., 2006). To facilitate the handling and distribution of the material at small scale, several methods have been developed that involve milling of the material to a very small particle size (Chundawat et al., 2008; Decker et al., 2009; Gomez et al., 2010; Song et al., 2010). A reduction in size, however, constitutes a mechanical pretreatment and thus can reduce the recalcitrance of the material improving the digestibility and thereby be able to mask possible differences among enzymes. To distribute the biomass in a reproducible and accurate manner, several alternative methods have been proposed, e.g., the direct pipetting of biomass slurry or dispensing fabricated handsheets of lignocellulosic material into the wells of micro-plates (Alvira et al., 2010b; Berlin et al., 2006;). The use of this assay is also more industrially relevant than the methods which use synthetic substrates such as filter paper and carboxymethyl cellulose. Additional advantages thereof include its potential to evaluate several factors with relative simplicity, its capability to screen cellulases and test enzyme synergism as well as its ability to discriminate between enzymatic preparations (Alvira et al., 2010b).

- Metabolic engineering of glycosylation pathways is considered a key element to develop improved eukaryotic cellulase producers.

- Reduction of the enzyme dosage required for efficient saccharification. Several strategies have been suggested to reduce the amount of enzyme required for complete enzymatic hydrolysis. These strategies can be grouped in terms of:
  - *Enhancement of the enzyme systems* by protein-engineering of known enzymes (mutagenesis, rational engineering, directed evolution) or by *bio-prospecting* of new enzymes (metagenomics and genomics approaches) (Kuhad et al., 2016) looking for advantageous properties:
    - Higher thermostability and specific activity.
    - Optimal binding capacity by removal/engineering the affinity of CBM according to substrate and process conditions. Although the presence of CBM facilitates the adsorption to crystalline cellulose, it also contributes to unproductive binding to lignin and cellulase obstruction. Enzymatic hydrolysis carried out at high solids content would not require the presence of CBM to promote the adsorption of the enzyme to cellulose (Várnai et al., 2013).
    - Low susceptibility to end-product inhibition, inhibitors generated during pretreatment and the ethanol from fermentation.
    - High tolerance to ionic liquids.
- *Process development*:
  - Use of *tailor-made enzyme combinations* adapted to feedstock and pretreatment (Pengilly et al., 2016). In this approach, combinations of cellulases with other enzyme activities or individual cellulase components and their ratios (Meyer et al., 2008) can be studied and optimized statistically for a particular biomass-pretreatment combination. The boost effect of enzymes in the appropriate ratio can lead to reduction of the final protein concentration needed for complete hydrolysis of the carbohydrates. Further, the incorporation on the hydrolysis media of non-catalytic additives such as surfactants, polymers and proteins has been proven to enhance enzyme performance thereby reducing the enzyme loading, as well as facilitating enzyme recycling (Jin et al., 2016).

    The biotechnological potential of "accessory enzymes" such as LPMOs needs to be exploited. More studies in terms of their diversity, role and mechanism of action on the degradation of crystalline cellulose are required (Martínez et al., 2016). Likewise, the incorporation of these enzymes in current commercial enzyme preparations must be evaluated for specific feedstocks (Hu et al., 2014), pretreatment (Rodríguez-Zúñiga et al., 2015) and process configurations (Canella & Jørgensen, 2014).

- *Process integration* of the biological steps:
  - Simultaneous saccharification and fermentation (SSF) alleviates end-product inhibition of cellulases, since the glucose generated from enzymatic hydrolysis are transformed into ethanol by the fermentative microorganism preventing their accumulation. The use of thermotolerant yeasts could overcome the different temperature optimum of enzymatic hydrolysis and fermentation (Tomás-Pejó et al., 2009).
  - Simultaneous saccharification and co-fermentation (SSCF) by development of microorganisms with the ability of fermenting not only hexoses (glucose, galactose, mannose) but also pentoses (xylose, arabinose) present in the pretreated materials. The SSCF in presence of accessory enzymes such as xylanases and xylosidase would prevent the accumulation of xylo-oligomers and xylose, which have been shown to impact negatively cellulose conversion.
  - Consolidated bioprocessing (CBP) configuration constitutes the maximum degree of integration, where a single microorganism or consortium of them is used for enzyme production, saccharification and co-fermentation in the same reactor (Den Haan et al., 2015). As a result, the externally added enzyme is reduced partially or totally.
  - Improvements in cellulose hydrolysis rates and ethanol productivity by optimising reaction conditions through process engineering. Other aspect that limits large scale cellulosic ethanol production is the use of high solids loading during enzymatic hydrolysis and/or fermentation. Solid concentrations higher than 15% of pretreated fibres are normally necessary to provide enough glucose that could yield this bench mark. However high solids loading entails problems of mixing and high viscosity that are being addressed by development of bioreactors with advanced mixing capacity and reduced energy consumption (Zhang et al.,

2010), by operating fed-batch processes (Cardona et al., 2015; Tomás-Pejó et al., 2009) and/or application of liquefaction step at higher temperatures prior to SSF (Jørgenssen et al., 2007).

CONCLUSION

Enzymatic hydrolysis of lignocellulosic biomass for biofuels production faces several challenges. Lignocellulosic materials are complex, heterogeneous and recalcitrant to cellulases and auxiliary enzymes. Although the application of a pretreatment prior to hydrolysis improves the digestibility, there is a fraction of the pretreated material difficult to hydrolyze, which result in an incomplete hydrolysis of the material. Thus, there is interest in the development of more efficient enzymes by manipulating desirable properties for the process such as thermostability and interaction with the substrate. The enzyme cocktail applied during enzymatic hydrolysis should be optimized considering the feedstock and the type and conditions of pretreatment.

Cost related to enzymes production and its use are one of the main obstacles that hamper the industrial expansion of cellulosic ethanol production. A better understanding of the mechanism of action of cellulases and accessory enzymes is crucial to improve the polysaccharide degradation into fermentable sugars. This will require a holistic approach of the ethanol production process considering all the steps and different disciplines such as genetics, biochemistry, biotechnology and process engineering.

REFERENCES

Alvira, P., Tomás-Pejó, E., Ballesteros, M. & Negro, M.J. (2010a). Pretreatment technologies for an efficient bioethanol production process based on enzymatic hydrolysis: A review. *Bioresource Technol*, 101, 4851–4861.

Alvira, P., Negro, M. J., Sáez, F., Ballesteros, M. (2010b). Application of micro-assay method to study the enzymatic hydrolysis of pretreated wheat straw. *J Chem Technol Biot*, 85, 1291-1297.

Arantes, V. & Saddler, J. N. (2011). Cellulose accessibility limits the effectiveness of minimum cellulase loading on the efficient hydrolysis of pretreated lignocellulosic substrates. *Biotechnol Biofuel* 4:3.

Banerjee, G., Scott-Craig, J. S., & Walton, J. D. (2010). Improving enzymes for biomass conversion: a basic research perspective. *Bioenerg Res*, 3(1), 82-92.

Bansal, P., Vowell, B. J., Hall, M., Realff, M. J., Lee, J. H., & Bommarius, A. S. (2012). Elucidation of cellulose accessibility, hydrolysability and reactivity as the major limitations in the enzymatic hydrolysis of cellulose. *Bioresource Technol*, 107, 243-250.

Bayer, E. A., Belaich, J. P., Shoham, Y. & Lamed, R. (2004). The cellulosomes: multienzyme machines for degradation of plant cell wall polysaccharides. *Annu Rev Microbiol*, 58, 521-554.

Bekham, G. T., Ståhlberg, J., Knott, B. C., Himmel, M. E., Crowley, M. F., Sandgren, M., Sørlie, M, & Payne, C. M. (2014). Towards a molecular-level theory of carbohydrate processivity in glycoside hydrolases. *Curr Opin Biotech*, 27, 96-106.

Benjamin, Y., García-Aparicio, M. P. & Görgens, J. F. (2014). Impact of cultivar selection and process optimization on ethanol yield from different varieties of sugarcane. Biotechnol Biofuels, 7(1), 60.

Berlin A., Maximenko V., Bura R., Kang K.-Y., Gilkes N. & Saddler J. (2006). A rapid microassay to evaluate enzymatic hydrolysis of lignocellulosic substrates. Biotechnol Bioeng, 93(5), 880–886.

Berlin, A, Gilkes, N., Kurabi, A., Bura, R., Tu, M., Kilburn, D. & Saddler, J. (2005). Weak lignin-binding enzymes: a novel approach to improve activity of cellulases for hydrolysis of lignocellulosics. *Appl Biochem Biotech*, 121 (1-3), 0163-0170, 2005.

Biofuels Digest: http://www.biofuelsdigest.com/ (Accessed June of 2016).

Cardona M. J., Tozzi E. J., Karunac N., Jeohc T., Powella R. L. & McCarthy M. J. (2015). A process for energy-efficient high-solids fed-batch enzymatic liquefaction of cellulosic biomass. *Bioresource Technol*, 198, 488–496.

CAZy database: http://www.cazy.org/Glycoside-Hydrolases.html.

Cherubuni, F. (2010). The biorefinery concept: Using biomass instead of oil for producing energy and chemicals. *Energ Convers Manage*, 51 (7), 1412-1421.

Conesa, A., Punt, P. J., van Luijk, N. & van den Hondel, C. A. M. J. (2001). The secretion pathway in filamentous fungi: a biotechnological view. Fungal Genet Biol, 33, 155-171.

Corrêa, T. L. R., Dos Santos, L. V. & Pereira, G. A. G. (2016). AA9 and AA10: from enigmatic to essential enzymes. *Appl Microb Biotech*, 100 (1), 9-16.

Cragg, S. M., Beckham, G. T., Bruce, N. C., Bugg, T. D. H., Distel, D. L., Dupree, P., Etxabe, A. G., Goodell, B. S., Jellison, J., McGeehan, J. E., Mc Queen-Mason, S. J., Schnorr, K., Walton, P. H., Watts, J. E. M. & Zimmer, M. (2015). Lignocellulose degradation mechanisms across the Tree of Life. *Curr Opin Chem Biol*, 29, 108-119.

Del Rio L. F., Chandra R. P. & Saddler J. N. (2012). Fiber size does not appear to influence the ease of enzymatic hydrolysis of organosolv-pretreated softwoods. Bioresource Technol, 107, 235-242.

Den Haan, R., van Rensburg, E., Rose, S. H., Görgens, J. F. & van Zyl, W. H. (2015). Progress and challenges in the engineering of non-cellulolytic microorganisms for consolidated bioprocessing. *Curr Opin Biotech*, 33, 15-22.

Dewan, S. S. (2013). EGY099A - Global markets and technologies for biofuel enzymes. BCC Market report. ISBN: 0-89336-222-0.

Dewan, S. S. (2014). BIO030H - Global Markets for Enzymes in Industrial Applications. BCC Market report.

Domingues, F. C., Queiroz, J. A., Cabral, J. M. S. & Fonseca, L. P. (2000). The influence of culture conditions on mycelial structure and cellulase production by *Trichoderma reesei*. Enzyme Microb Tech, 26 (5–6), 394–401.

Druzhinina, I. S. & Kubicek, C. P. (2016). Familiar Stranger: Ecological genomics of the model saprotroph and industrial enzyme producer *Trichoderma reesei* breaks the stereotypes. Adv *Appl Microbiol*, 95, 69-147.

Eckborg, N. A., Morrill, W., Burgoyne, A. M., Li, L. & Distell, D. L. (2007). CelAB a multifunctional cellulase encoded by *Teredinibacter turnerae* T7902(T), a culturable symbiont isolated from the wood-boring marine bivalve *Lyrodus pedicellatus*. *Appl Environ Microbiol*, 73, 7785-7788.

Elgharbawy, A. A., Alam, M. Z., Moniruzzaman, M. & Goto, M. (2016). Ionic liquid pretreatment as emerging approaches for enhanced enzymatic hydrolysis of lignocellulosic biomass. *Biochem Eng J*, 109, 252-267.

García-Aparicio, M. P., Ballesteros, M., Manzanares. P., Ballesteros, I., González, A. & Negro, J. M. (2007). Xylanase contribution to the efficiency of cellulose enzymatic hydrolysis of barley straw. *Appl Biochem Biotechnol*, 137-140 (1-12), 353-365.

Gomes, J., & Steiner, W. (2004). The biocatatlytic potential of extremophiles and extremozymes. *Food Tecnol Biotechnol*, 42(4), 223-235.

Gyalai-Korpos, Nagy, G., Mareczky, Z., Schuster, A., Réczey, K., & Schmoll, M. (2010). Relevance of the light signalling machinery for cellulase expression in *Trichoderma reesei* (*Hypocrea jecorina*). *BMC Research Notes*, 3, 330.

Henriksson, H., Stahlberg, J., Isaksson, R. & Pettersson, G. (1996). The active sites of cellulases are involved in chiral recognition: A comparison of cellobiohydrolase I and endoglucanase I. *FEBS Letters*, 390, 339-344.

Horn, S. J.S. J., Vaaje-Kolstad, G., Westereng, B. & Eijsink, V. G. (2012). Novel enzymes for the degradation of cellulose. *Biotechnol Biofuels*, 5(1), 45.

http://www.cazy.org/Carbohydrate-Binding-Modules.html.

https://www.cazypedia.org/index.php/Main_Page.

Hu, J., Arantes, V. & Saddler, J. N. (2011). The enhancement of enzymatic hydrolysis of lignocellulosic substrates by the addition of accessory enzymes such as xylanase: is it an additive or synergistic effect? *Biotechnol Biofuels,* 4, 36.

Imman, S., H., Bilbao-Sainz, C., Chiou, B.-S., Glenn, G. M., Orts, & William J. (2013). Biobased adhesives, gums, emulsions, and binders: current trends and future prospects. *J Adhes Sci Technol*, 27 (18-19), 1972–1997.

Jin, W., Chen, L., Hu, M., Sun, D., Li, A., Li, Y., Hu, Z., Zhou, S., Tu, Y., Xia, T., Wang, Y., Xie, G., Li, Y, Bai, B., & Peng, L. (2016). Tween-80 is effective for enhancing steam-exploded biomass enzymatic saccharification and ethanol production by specifically lessening cellulase absorption with lignin in common reed. *Appl Energ*, 175, 82-90.

Johnson, E. (2016). Integrated enzyme production lowers the cost of cellulosic ethanol. Biofuels, Bioprod Biorefin, 10(2), 164-174.

Jørgensen, H. & Olsson, L. (2006). Production of cellulases by *Penicillium brasilianum* IBT 20888 - Effect of substrate on hydrolytic performance. *Enzyme Microb Tech*, 38, 381-390.

Jørgensen, H., Kristensen, J. B., & Felby, C. (2007). Enzymatic conversion of lignocellulose into fermentable sugars: challenges and opportunities. *Biofuels, Bioprod Biorefin*, 1(2), 119-134.

Juhász, T., Szengyel, Z., Reczey, K., Siika-Aho, M. & Viikari, L. (2005). Characterization of cellulases and hemicellulases produced by *Trichoderma reesei* on various carbon sources. *Process Biochem*, 40, 3519-3525.

Klein-Marcuschamer D., Oleskowicz-Popiel P., Simmons B. A. & Blanch H. W. (2012). The challenge of enzyme cost in the production of lignocellulosic biofuels. *Biotechnol Bioeng*, 109(4), 1083-1087.

Kuhad, R. C., Deswal, D., Sharma, S., Bhattacharya, A., Jain, K. Ka, Kaur, A., Pletschke, B. I., Singh, A. & Karp, M. (2016). Revisiting cellulase production and redefining current strategies based on major challenges. *Renew Sust Energ Rev*, 55, 249-272.

Lawson, K. (2011). Global markets for marine-derived pharmaceuticals. Bcc market report PHM101A.

Lemos, M. A., Teixeira, J. A., Domingues, M. R. M., Mota, M. & Gama, F. M. (2003). The enhancement of the cellulolytic activity of cellobiohydrolase I and endoglucanase by the addition of cellulose binding domains derived from *Trichoderma reesei. Enzime Microb Technol*, 32, 35-40.

Levasseur, A., Drula, E., Lombard, V,. Coutinho, P. M., & Henrissat, B. (2013) Expansion of the enzymatic repertoire of the CAZy database to integrate auxiliary redox enzymes. *Biotechnol Biofuels*, 6:41.

Lo, C. M., Zhang, Q., Lee, P., & Ju, L. K. (2005). Cellulase production by *Trichoderma reesei* using sawdust hydrolysate. *Appl Biochem Biotechnol*, 121-124, 561-573.

Martínez, A. T. (2016). How to break down crystalline cellulose. Science 352: 6289.

Medve, J., Karlsson, J., Lee, D. & Tjerneld, F. (1998). Hydrolysis of microcrystalline cellulose by cellobiohydrolase I and endoglucanase II from *Trichoderma reesei*: Adsorption, sugar production pattern, and synergism of the enzymes. *Biotechnol Bioeng*, 59, 621-634.

Medve, J., Karlsson, J., Lee, D. y Tjerneld, F. (1998). Hydrolysis of microcrystalline cellulose by cellobiohydrolase I and endoglucanase II from *Trichoderma reesei*: Adsorption, sugar production pattern, and synergism of the enzymes. *Biotechnol Bioeng*, 59, 621-634.

Meyer, A. S., Rosgaard, L. & Sørensen, H. R. (2009). The minimal enzyme cocktail concept for biomass processing. *J Cereal Sci*, 50(3), 337-344.

Newman, R. H., Vaidya, A. A., Sohel, M. I. & Jack, M. W. (2013). Optimizing the enzyme loading and incubation time in enzymatic hydrolysis of lignocellulosic substrates. *Bioresour Technol,* 21, 1075-1080.

OECD 2001. The Application of Biotechnology to Industrial Sustainability (www.oecd.org/sti/biotechnology).

Olofsson, K., Bertilsson, M., & Lidén, G. (2008). A short review on SSF – an interesting process option for ethanol production from lignocellulosic feedstocks. *Biotechnol Biofuels*, 1(1), 7.

Olsson, L., Jorgensen, H., Krogh, K. B. R. & Roca, C. (2005). Bioethanol production from lignocellulosic material. In: Polysaccharides structural diversity and functional versatility. S. Dimitriu (Ed), Marcel Dekker, New York, 957-993.

Pengilly, C., García-Aparicio, M. P., Diedericks, D. & Görgens, J. F. (2016). Optimization of enzymatic hydrolysis of steam pretreated triticale straw. *Bioenerg Res*, 1-13.

Pengilly, C., García-Aparicio, M. P., Diedericks, D., Brienzo, M. & Görgens, J. F. (2015). Enzymatic hydrolysis of steam-pretreated sweet sorghum bagasse by combinations of cellulase and endo-xylanase. *Fuel*, 154, 352-360.

Persson, I., Tjerneld, F., Hahn-Hagerdal, B. (1991). Fungal cellulolytic enzyme production: A review. *Process Biochem*, 26, 65-74.

Quentin, M., van der Valk, H., van Dam, J. & de Jong, E. (2003). Cellulose-binding domains: Tools for innovation in cellulosic fiber production and modification. Applications of enzymes to lignocellulosics, ACS Symposium Series, 855, 132-155.

Rabinovich, M. L., Melnick, M. S. & Bolobova, A. V. (2002). The structure and mechanism of action of cellulolytic enzymes. *Biochemistry-Moscow*, 67, 850-871.

Rodríguez-Zúñiga, U. F., Cannella, D., Giordano, R. D. C., Giordano R. D. L. C., Jørgensen, H., & Felby, C. (2015). Lignocellulose pretreatment technologies affect the level of enzymatic cellulose oxidation by LPMO. *Green Chem*, 17(5), 2896-2903.

Sánchez, Ó. J. & Cardona, C. A. (2008). Trends in biotechnological production of fuel ethanol from different feedstocks. *Bioresource Technol*, 99, 5270–5295.

Schomburg, D., & Schomburg, I. (2010). Enzyme database. Methods in molecular biology (Clifton, N. J.) 609, 113-128.

Sheng, J., Agblevor, F. A. (2008). Optimization of enzyme loading and hydrolytic time in the hydrolysis of cotton gin waste and recycled paper sludge for the maximum profit rate. *Biochem Eng J*, 41, 241-250.

Tengerdy, R. P. & Szakacs, G. (2003). Bioconversion of lignocellulose in solid substrate fermentation. *Biochem Eng J*, 13, 169-179.

Thongekkaew, J., Ikeda, H., Masaki, K., & Iefuji, H. (2013). Fusion of cellulose binding domain from *Trichoderma reesei* CBHI to *Cryptococcus* sp. S-2 cellulase enhances its binding affinity and its cellulolytic activity to insoluble cellulosic substrates. *Enzyme Microb Technol*, 52(4-5), 241-246.

Tomás-Pejó, E., Oliva, J. M., González, A., Ballesteros, I. & Ballesteros, M. (2009). Bioethanol production from wheat straw by the thermotolerant yeast *Kluyveromyces marxianus* CECT 10875 in a simultaneous saccharification and fermentation fed-batch process. *Fuel*, 88, 2142-2147.

Várnai, A., Mäkelä, M. R., Djajadi, D. T., Rahikainen, J., Hatakka, A., & Viikari, L. (2014). Carbohydrate-binding modules of fungal cellulases. Occurrence in nature, function, and relevance in industrial biomass conversion. *Adv Appl Microbiol*, 88, 103-165.

Várnai, A., Siika-aho, M., & Viikari, L. (2013). Carbohydrate-binding modules (CBMs) revisited: reduced amount of water counterbalances the need of CBMs. *Biotechnol Biofuels*, 6, 30.

Viikari, L., Alapuranen, M., Puranen, T., Vehmaanpera, J. & Siika-Aho, M. (2007). Thermostable enzymes in lignocellulose hydrolysis. *Adv Biochem Eng Biotechnol*, 108, 121-45.

Viikari, L., Vehmaanperä J., & Koivula, A. (2012). Lignocellulose ethanol: from science to industry. Biomass Bioenerg, 46, 13-24.

Vuong, T. V., & Wilson D. B. (2010). Glycoside Hydrolases: Catalytic base/nucleophile diversity. Biotechnol Bioeng, 107(2), 195-205.

Wallace, J., Brienzo, M., García-Aparicio, M. P. & Görgens, J. F. (2016). Lignin enrichment and enzyme deactivation as the root cause of enzymatic hydrolysis slowdown of steam pretreated sugarcane bagasse. *New Biotechnol*, 33(3), 361-371.

Walker, J. A., Takasuka, T. E., Deng, K., Bianchetti, C. M., Udell, H. S. & Prom, B. M. (2015). Multifunctional cellulase catalysis targeted by fusion to different carbohydrate-binding modules. *Biotechnol Biofuels*, 8(1), 220.

Wang, N., Zang, J., Ming, K., Liu, Y., Wu, Z., & Ding, H. (2013). Production of cold-adapted cellulase by Verticillium sp. isolated from Antarctic soils. *Electron J Biotechnol*, 16(4).

Xiao, Z. Z., Zhang, X., Gregg, D. J. & Saddler, J. N. (2004). Effects of sugar inhibition on cellulases and beta-glucosidase during enzymatic hydrolysis of softwood substrates. *Appl Biochem Biotechnol*, 113-16, 1115-1126.

Yue, Z., Bin, W. & Peiji, G. (2004). Mechanism of cellobiose inhibition in cellulose hydrolysis by cellobiohydrolase. Science in China. Series C, Life sciences/Chinese Academy of Sciences, 47, 18-24.

Zhang, J., Chu, D., Huang, J., YU, Z., Dai, G., Ban J. (2010). Simultaneous Saccharification and ethanol fermentation at high corn stover solids loading in helical stirring bioreactor. *Biotechnol Bioeng*, 105, 718-728.

Zhang, Y. H. P. (2011). What is vital (and not vital) to advance economically-competitive biofuels production. *Process Biochem*, 46, 715-725.

In: Cellulase
Editor: Marguerite Stevenson
ISBN: 978-1-63485-788-8

Chapter 4

CELLULASE CHALLENGES OF LIGNOCELLULOSIC MATERIAL HYDROLYSIS

Michel Brienzo[1,*], María García-Aparicio[2,3], Luvuyo Tyhoda[4], Celso Sant'Anna[5] and Johann Görgens[3]

[1]Bioenergy Research Institute (IPBEN),
Universidade Estadual Paulista (UNESP), Rio Claro-SP, Brazil
Departments of [2]Microbiology,
[3]Process Engineering, and
[4]Forest and Wood Science, Stellenbosch University, South Africa
[5]Laboratory of Biotechnology (Labio),
Metrology Applied to Life Science Division -
National Institute of Metrology, Quality and Technology (Inmetro),
Duque de Caxias - RJ, Brazil

ABSTRACT

Cellulases comprise a group of glycoside hydrolytic enzymes that cleave β-1,4-glycosidic linkages of cellulose into glucose units. These enzymes have been applied industrially for more than three decades. They have found application in different industrial sectors namely textile, detergents, pulp and paper, food, beverages and the feed industry.

---

* Corresponding author: mbrienzo@ipben.unesp.br.

Notwithstanding, research in cellulase enzymes and their mechanism of action has increased steadily owing to the interest in the production of biofuels such as ethanol from lignocellulose materials. The biological conversion of cellulosic material is a sustainable approach focused on waste reduction and valorization. However, cellulose in lignocellulosic materials is highly organized and requires that the biomass must be pretreated and enzymatically hydrolyzed to release the fermentable sugars in appreciable amounts. Cellulose is associated with hemicelluloses and lignin in the rigid plant cell wall structures, which are naturally resistant to microbial and enzymatic degradation. The enzymatic hydrolysis of cellulose faces technical challenges related to both the nature of the substrate and enzyme factors. Despite the additional pretreatment step, there is normally a fraction of cellulose that remains un-hydrolyzed after enzymatic hydrolysis. Cellulases also display reduced stability under industrial application conditions such as temperature, the presence of inhibitors and end-product inhibition. This chapter addresses substrate factors related to enzymatic hydrolysis of cellulose and lignocellulosic materials. The cellulose and lignocellulosic materials are reviewed based on their chemical composition and physicochemical properties. Furthermore, these properties are related to the challenge of cellulases hydrolysis of cellulose and cellulosic materials. The effective hydrolysis of cellulose would allow for economically feasible production of biofuels (ethanol, hydrogen), chemicals and polymers.

INTRODUCTION

Lignocellulose biorefineries are envisioned as key element in the bioeconomy strategy for the production of biofuels, chemicals and biomaterials from renewable sources. Lignocellulose biomass derives from plant cell walls and is comprised mainly by cellulose, hemicelluloses and lignin. Although the effective exploitation of these three fractions is considered crucial for the economic viability of lignocellulose biorefineries, current technology has been mainly focused on cellulose utilization for its conversion into ethanol, so called second generation ethanol (2G). 2G ethanol production consists of several steps including reduction of particle size, application of a pretreatment, production and application of enzymes (cellulases, hemicellulases and other accessory enzymes), fermentation of sugars (hexoses and pentoses) and distillation.

The complexity of lignocellulosic biomass requires a large enzymatic pool to deconstruct the cell wall and release fermentable sugars. Cellulases and hemicellulases (xylanases and mannanases) degrade carbohydrates and ligninases (laccases, lignin peroxidase and manganese peroxidase, in association with low-molecular weight mediators) degrade lignin. However, lignocellulose is characterized by a complex design with diverse structural sizes (from Ångström to micrometres) (Chundawat et al., 2011; Langan et al., 2011). These complexity is responsible of its recalcitrance, that is its resistance to microbial and enzymatic degradation. Several pretreatments have been developed to reduce the lignocellulose recalcitrance and thus improve the accessibility of cellulose to enzymes (Alvira et al., 2010).

Despite the advances of pretreatment and enzyme technology, the enzymatic saccharification of polysaccharides present in lignocelluloses constitutes a limiting step for its conversion into biofuels and chemicals. This is owing to the high cost of production of cellulases and the high dosages required to achieve relatively good conversion yields. Furthermore, the enzymatic hydrolysis rate falls sharply during the process, what adds to the enzyme requirements. This reduction of conversion rate is observed in both pure cellulose substrates (i.e., Avicel) and pretreated lignocelluloses substrates. The hydrolysis is influenced by factors related to enzyme characteristics, which include the nature and source of the enzyme. The challenge with regards to substrate is mainly its recalcitrance which is influenced by the substrate (chemical and physical properties). When using lignocellulose as substrate, the enzymatic hydrolysis rates and conversion will be determined by feedstock (softwood, hardwood, herbaceous), fraction of the feedstock and type and conditions of pretreatment applied.

The accessibility of the cellulose to the cellulolytic system plays a major role in the enzymatic hydrolysis. Cellulose accessibility is determined by the type of feedstock and pretreatment that is applied. Cellulose presents both external surface (size and shape of the fibres) and internal surface (capillar structure of the fibres). Both surfaces are more accessible to cellulases after it is subjected to a pretreatment step that generally remove the hemicelluloses while altering the lignin. The reactivity of the substrate, that is its susceptibility to enzyme action, relies on several substrate properties concerning the cellulose, hemicellulose, lignin and other minor components. The degree of polymerization of cellulose will determine the amount of reducing and non-reducing ends available to exoglucanases. Another structural property of the cellulose, the crystallinity, also influences the degree of conversion; although its detrimental effect on hydrolysis rate is mainly related to loss of synergism among the components of the cellulolytic system. The crystalline nature of cellulose depends on the proportion of cellulose allophormes that coexist in the native cellulose, that in turn determine the hydrogen-bonding pattern and thus its digestibility. Furthermore, cellulose is associated with hemicelluloses and lignin, forming the plant cell wall (Perez and Mazeau, 2004; De Souza and Sant'Anna, 2012). The cellulose fibers are protected by a network of polysaccharides of hemicelluloses and encrusted by lignin (Sant'Anna et al., 2014). Hemicelluloses and lignin act as physical barriers to the enzymes. The properties of hemicelluloses, such as degree of substitution and acetylation, have been shown to influence their recalcitrance (Benjamin et al., 2014). In terms of lignin, the percentage of functional groups will determine the events of non-productive adsorption of cellulases (Zhao et al., 2012). Furthermore, lignin and polysaccharides, mainly hemicelluloses, are covantly linked originating lignin-carbohydrate complexes (LCCs). It has been suggested that the (LCCs) functions also as a barrier protecting "cellulose-like" hemicelluloses from enzymatic depolymerization (Min et al., 2014).

Cellulose as substrate, whether it is in isolated form or part of the cell walls of lignocellulosic materials, directly influences the enzymatic hydrolysis process in terms of fermentable glucose release rate and yield. Therefore, to break the cellulose free or at least increase its accessibility, the lignocellulosic matrix needs a pretreatment step to modify its properties such as porosity and expose the cellulose to enzyme action. The degree of lignocellulosic matrix organization and recalcitrance is complex and as such it is difficult to assess which cellulose physicochemical properties are more important for an efficient enzymatic hydrolysis process. The efficiency of the enzymatic hydrolysis process for either pure or isolated cellulose can be correlated to some of the physicochemical properties as crystallinity and DP. However, this correlation is not directly related to the enzymatic hydrolysis of the gross lignocellulosic substrate. As such, the cellulose or the gross material properties should be evaluated together, since the pretreatment affects all of them. Similarly, the biomass response to the pretreatment in terms of degree of modification or improvement on the processability is in turn related to the its chemical composition and physicochemical properties, both of which vary across different materials and sometimes different varieties of the same material (Brienzo et al., 2015; Benjamin et al., 2013; Masarin et al., 2011). Moreover, the biomass is heterogenic, which strongly contributes to its recalcitrance (Brienzo et al., 2016, 2014; Sant'Anna et al., 2014; Costa et al., 2013; Chundawat et al., 2011).

The combination of these factors result in end-product inhibition, enzyme thermal inactivation, unproductive binding or blockage of the cellulases in internal pores of cellulose and reduction of amorphous region of cellulose (more amenable to enzyme action) which will in turn lead to slowdown in the cellulose conversion. There is no consensus in what factors are more relevant in the hydrolysis rates but they are generally grouped in substrate-related and those related to the source and mechanism of action of the cellulase system employed, including reaction conditions. A better understanding of the role of these factors will allow for the development of feedstocks with desired properties, novel pretreatment technologies, superior enzyme catalysts and process configurations with improved saccharification yields, and thus, higher overall product yield. The following sections will describe the substrate related factors as physicochemical and structural features of the biomass.

SUBSTRATE-RELATED FACTORS

There are several factors responsible for the reduction of the hydrolysis rate and glucose yield during lignocellulosic material versus pure cellulose bioconversion. These factors can be divided into substrate- and enzyme-related factors as shown in Table 1. Some of these factors are strictly dependent on the cellulose characteristics and others are related to the overall structure and nature of the lignocellulosic material. Similarly, enzyme related factors are intrinsically related to cellulases properties, and reaction conditions. Currently, there is no consensus as to which factors are more important. However, enzyme- or substrate-related factors could be related to the type of biomass, pretreatment conditions and enzyme extract applied (Del Rio et al., 2012).

The rates and yields of enzymatic hydrolysis are dependent on the substrate structure and composition. Lignocellulosic substrates show different recalcitrance during enzymatic hydrolysis (Fan et al., 1981). The substrate origin or composition, the pretreatment method applied and its condition are certain to result in significant differences in the material and enzyme interaction, and consequently the resultant glucose yield. The substrate related factors include cuticle and waxes in the epidermal tissue of the plant, vascular bundle arrangement and density, degree of lignification and phenolics, the wall thickness of the sclerenchyma tissue, microfibril and matrix polymer heterogeneity (acetyl and hemicelluloses), insoluble substrates, crystallinity and the degree of polymerization (DP) of cellulose (Brienzo et al., 2016; Hallac and Ragauskas, 2011; Himmel et al., 2007; Zhao et al., 2012). Crystallinity and DP are specific properties of cellulose which are different for pure cellulose and cellulose in the lignocellulosic material. These two important physicochemical properties will be discussed separately. The recalcitrance related to the material heterogeneity and complexity will also be discussed.

Lignin is one of the cell wall components and affects the accessibility of cellulose to cellulase by acting as a physical barrier for cellulase accessibility and adsorption of enzymes which reduces their effectiveness and productivity. Accordingly, the content as well as the type of lignin have impacts on the hydrolysis of lignocellulosic materials (Mooney et al., 1998). Lignin distribution based on the tissue and type of cell, i.e. vessels, parenchyma and sclerenchyma (Brienzo et al., 2016), and the lignin composition (Guo et al., 2014) can contribute to the recalcitrance of the lignocellulose materials. The composition of the lignin which is discussed in detail later has an influence in the recalcitrance of biomass and contributes to enzyme adsorption during the hydrolysis. The removal of lignin to facilitate hydrolysis has several advantages in that the available cellulose surface area is increased and the, counterproductive enzyme binding is reduced, both of which contribute to an increase cellulose hydrolysis efficiency (Wu et al., 2011).

The hemicelluloses, as plant cell wall components also impose a barrier for enzyme access to the cellulose and thereby reduce the efficiency of hydrolysis. Therefore, their modification or removal is paramount in opening up the biomass structure to ease the access of the enzymes to the cellulose. Hemicelluloses form a network surrounding the cellulose fibres (Sant'Anna et al., 2014). For pretreated material with residual hemicelluloses, the employment of cellulases combined with hemicellulases in the hydrolysis step results in an increase in cellulose digestibility and also achieve synergistic benefits (Berlin et al., 2006). The residual acetyl groups of the hemicelluloses which are found in untreated and pretreated materials interfere with the recognition of enzymes due steric hindrance, which decreases the access of cellulase (Waldron, 2010). Hemicelluloses and lignin, which are substrate related inhibitors have been suggested to be the most important physicochemical properties influencing the enzymatic hydrolysis, and hence their modification or removal is paramount in order to increase cellulose accessibility.

For industrial application, the concentration of substrate or solids loading used in the enzymatic hydrolysis process is an important factor. Hydrolysis at high solid loading leads to an increased glucose yield and consequently high ethanol yield, which reduces the cost of product recovery due to the requirements for distillation (Fierobe et al., 2002). However, at high solids loading, several process problems can occur, such as those related to cellulase effectiveness (sugar inhibition) and sufficient substrate mixing (Fierobe et al.,

2002). Viscosity therefore has impact on enzyme-substrate due mixing difficulties, and should be decreased.

Table 1. Enzyme and substrate related factors influence on the rate and yield of enzymatic hydrolysis of lignocellulosic material

Substrate factors	Enzyme factor
Crystallinity*	Thermal stability#
Degree of polymerization*	Mechanical stability#
Structural heterogeneity	Synergism#
Lignin content	Product inhibition#
Lignin composition	Reaction condition: pH, Temperature, agitation and time
Hemicelluloses content	Tolerate degradation products from sugars and lignin
Pore distribution	
Pore size and volume	
Specific surface area	
Particle size	

(*) intrinsic properties of the cellulose; (#) intrinsic properties of the cellulases.

Table 2. The limiting factors related to the biomass recalcitrance at different scale

Biomass recalcitrance	Factors related
Macro-scale	The plant anatomy, tissue organization, and cell diversity
Micro-scale	The chemical composition, lignin concentration and localization, cell wall thickness, and lignin sealing of the cellulose and hemicellulose sheathing
Nano-scale	The crystallinity of the cellulose, degree of polymerization and the cell wall porosity

PHYSICOCHEMICAL PROPERTIES

The lignocellulosic material is constituted of different cell types whose cell wall organization at the macro-, micro-, and nanoscales contributes to its structural heterogeneity and recalcitrance (Table 2) (Sant'Anna et al., 2014). The recalcitrance, due this higher organization, increases biomass resistance to breaking it down to its main structural components i.e., lignin, hemicelluloses and cellulose, of which the latter is a source of fermentable glucose (Lynd et al., 2008). The lignocellulosic material properties with an intrinsic influence on its recalcitrance include biomass particle size, cell wall thickness, cell wall porosity, cellulose accessibility, cellulose crystallinity, cellulose degree of polymerization and lignin content and its distribution.

The cellulose properties within biomass samples vary according to the plant cell wall organization, which also varies according to biomass species and varieties within a species, maturation or age of the sample, presence of defects such a reaction wood (in the case of tension wood there is more cellulose and less lignin in the biomass, while the reverse is true for compression wood), as well as location in the biomass. Studies have shown that in woody biomass, wood properties such as density follow the maturation of the biomass and increase from pith to bark and from top to bottom of the stem. There are however exceptions to this trend with regards to ring porous hardwood species, which could be exacerbated even further under conditions which promote slow growth. Cellulose like density proportionally increases from pith to bark following the maturation of the wood (with exception of ring porous hardwoods as discussed above), while lignin follows an opposite trend and progressively decreases (Ververis et al., 2004). Due these factors, the pretreatment process employed to alter the structure of the biomass must be tailored for the specific biomass sample. In sugarcane, the culm can be also separated into an internode, node and external fraction (which contains epidermis), that respectively increases the recalcitrance to acid pretreatment (Brienzo et al., 2014). The number, the size and distribution of vascular bundles were found to contribute to differences in recalcitrance in sugarcane (Brienzo et al., 2016). The recalcitrance of the sugarcane culm also increases from the center to periphery (Costa et al., 2013). The vascular bundles in grasses such as sugarcane have secondary helical thickenings of 2.5 mm wide, surrounded by a thick layer of highly lignified sclerenchyma cells. The sclerenchyma cells provide mechanical support and protective effect on the tissue structure (Sant'Anna et al., 2013). The characteristics and properties of the vascular bundles are not totally understood. Therefore, work still lies ahead in terms of establishing the constitution of vascular bundles and its contribution to recalcitrance. Similarly, the recalcitrance of corn stover depends on the location of the tissue within the biomass. The center part of the corn stover stalk consists primarily of parenchyma cells, and the external fraction comprises of sclerenchyma cells, vascular tissue and epidermal cells (Zeng et al., 2012).

The lumen size, cell wall thickness and the overall diameter of the xylem elements such as vessels, fibres, tracheids, parenchyma cells, sclerenchyma cells and their relative proportions in the biomass could be related to the degree of the recalcitrance of the biomass. The sclerenchyma cell wall thickness contribute to recalcitrance in addition to the lignin content, since the walls are highly lignified. Besides the structural features, the lignin content has been considered as important factor contributing to biomass recalcitrance. For enzyme action to be effective on the cellulose in the lignocellulosic material, a pretreatment step is necessary to change its structure and make cellulose accessible. The pretreatment effectiveness depends on the material type, such as hardwood, softwood and grasses in relation to the properties discussed above. The fraction or tissue location in the biomass was also found to play a major role in the enzymatic hydrolysis yield with the highest yield observed in the internode than both the node and external fraction (Brienzo et al., 2014). The digestibility of the sugarcane culm (internode) decreases from the center to the periphery (Costa et al., 2013). Therefore, each respective fraction requires its own optimum pretreatment conditions to in order to change the chemical and physicochemical properties of the biomass and improve glucose yield during enzymatic hydrolysis.

Cellulose has a linear homopolymeric structure composed of D-glucopyranose units linked by β-1,4 bonds. The cellulose chemical formula is $C_6H_{10}O_5)n$, and is composed of 44.44% carbon, 6.17%, hydrogen (and 49.39%. However, its content in biomass varies among the various plants such as hardwood versus softwood or grasses, in the range between 40-50%, 40-45% and 25-50%, respectively (Howard et al., 2003; Sjostrom, 1993; Fengel and Wegner, 1984). Furthermore, different types of cellulose are generated according to origin, morphology and extraction process. Celluloses are also differentiated in their physical structure, based on the spatial arrangement, aggregation molecules and crystal structure (allomorphs structures such as cellulose I and II). These properties influence the solubility of cellulose in most of the solvents, and also it's swelling ability. For example, cellulose is not soluble in water and is not affected by the temperature (Zhan, 2005). These properties have a direct impact on the efficiency of the enzymes in that they determine the enzymatic hydrolysis rate and glucose yield.

Although the mechanical properties are not the focus of this chapter, they however should not be ignored as they have a direct impact on wood performance in various applications such as natural fiber application, composite reinforcing fibers and applications of cellulose such as cellulose nanofibers or nanomaterials. Physical properties have a direct influence on the effectiveness of biomass enzymatic hydrolysis, hence pretreatment is required to deconstruct the biomass structure. Some of the most important mechanical properties of fibers include density, tensile strength and elongation at break. The mechanical properties of cellulosic materials have been recently reviewed at micro and nanoscale levels, wherein different materials and wood species were compared (Wang et al., 2010).

The factors responsible for the biomass recalcitrance are not independent and their individual effects cannot be isolated for investigation. Furthermore, these factors are multiple and interrelated, and pretreatment cannot change them individually. The following discussion will focus on the main biomass characteristics features that affect the recalcitrance of isolated or pure cellulose, and cellulose in lignocellulosic materials.

Crystallinity and Crystallinity Index

The organization of cellulose microfibrils is based on hydrogen bonds linking the cellulose molecules at intra- and inter-chain level, resulting in the formation of a supramolecular structure with different organization levels. These regions can be highly organized crystalline, or loosely organized amorphous regions. In general, the cellulose microfibrils are accepted to be largely crystalline, and the amorphous regions characterize surface imperfections. Native cellulose is comprised of a polymorphic structure commonly defined as cellulose I, which is found in the form of two crystalline phases known as I_α and I_β (Atalla and Vander Hart, 1984). Chemical treatments of cellulose I can then convert it to polymorphs II, III, and IV. The cellulose polymorphs have intrinsic characteristic x-ray diffraction patterns, which allow for their specific identification. The most abundant cellulose form is Cellulose I or the so called native cellulose. Cellulose I can be irreversibly modified to a more stable crystalline form, called cellulose II, or "regenerated cellulose," through regeneration and mercerization processes (Atalla and Vander Hart, 1984). The natural distribution of cellulose Iα and Iβ depend on the biomass type. Cellulose Iα is predominantly found in organisms such as bacteria and algae. Cellulose Iβ is predominantly found in higher plants such as woody biomass and cotton. Conformations, structures and morphologies of different cellulose types have been reviewed, with focus on detailing the polymorphic characteristics (Pérez and Mazeau, 2005).

The properties of the cellulose depend on the interactions that occur at the surface of the fibers and in between the chains. The chains can form dense structures in which the hydroxyl groups are involved in a crystal and cohesive structure. This highly organized structure has a different reactivity compared to the hydrogen bond involved in the interaction process, such as with adsorption with other molecule of cellulose or other cell wall macromolecules. The exposure of the hydroxyl groups or their reactivity can be defined at the fiber surface level or internally within the fibre. The crystallinity of the cellulose depends on such factors as the nature of the crystal lattice, proportion of crystalline region, and the size and orientation of the crystals (Ward, 1950). The organization of the cellulose in the crystalline region reduces the accessibility for enzymatic digestion compared to amorphous region; which facilitates the contact between enzymes and cellulose molecules, and thereby promoting better efficiency of enzyme action. The proportion of the crystalline region in a lignocellulosic material has a direct effect on its digestibility (Yoshida et al., 2008). The amorphous fraction can be hydrolyzed five to ten times faster than crystalline region. The initial rate of the enzymatic hydrolysis therefore is also dependent of the amount of crystalline or amorphous region present in the material (Laureano-Perez et al., 2005).

Cellulose crystallinity is of interest since it affects the enzymatic hydrolysis and consequently its conversion through biotechnological routes. The crystalline fraction in the cellulose is measured as the crystallinity index. This crystallinity index refers to the crystalline fraction of a lignocellulosic biomass, since lignin and hemicelluloses contribute to the amorphous fraction of the material (Wallace et al., 2016; Brienzo et al., 2016; Sant'Anna et al., 2014). In this regard, the reference is not directed to the crystalline cellulose fraction. Therefore, to differentiate the crystallinity index of cellulose from that of the lignocellulosic material, it is critical to understand the pretreatment effect and enzymatic hydrolysis. As mentioned earlier, cellulose contains a fraction defined as crystalline and another defined as amorphous. However, considering lignocellulosic materials, there are crystalline (cellulose crystalline fraction contribution) and amorphous (cellulose amorphous fraction, hemicellulose and lignin contribution) fractions. Pretreatment may affect the biomass by removing hemicelluloses or lignin, thus contributing to a decrease in the amorphous components, resulting in a proportional increase in the amount of crystalline fraction, which in turn increases the crystallinity index. Pretreatments such as the acid pretreatment, which partially removes hemicelluloses result in an increase in the crystallinity index of the pretreated material (Brienzo et al., 2014, 2015, 2016). The enzymatic hydrolysis of cellulose in biomass is specific and results in the removal of cellulose and thereby decreasing the amount of crystalline fraction in the material, which in turn decreases the crystallinity index of the material (Wallace et al., 2016).

The enzymatic hydrolysis of cellulosic material such as Avicel, on the other hand requires a different interpretation with regards to the crystallinity index, since there is no influence of non-cellulose components on the measurement of crystallinity. Several works have suggested that the crystallinity of the cellulose has a major role on the enzymatic hydrolysis rate. It is accepted that the amorphous fraction is hydrolyzed faster than the crystalline (Fan et al, 1980; Laureano-Perez et al., 2005; Zhang and Lynd, 2004). Considering that during the course of enzymatic hydrolysis the amorphous part of the cellulose is hydrolyzed first, the unhydrolyzed part becomes more recalcitrant due to the proportional increase in its crystallinity index. The increase in the cellulose recalcitrance by enrichment in crystalline fraction is directly related to a decrease in the enzymatic hydrolysis rate (Chang and Holtzapple, 2000; Hall et al., 2010). Previous works have suggested that the crystallinity index could be directly responsible for biomass recalcitrance, and postulated that it could then be used to predict the biomass digestibility. However, crystallinity alone cannot be used to explain the resistance of biomass to enzymatic hydrolysis. In fact, enzymatic hydrolysis is more influenced by the conjunct of the material physicochemical properties (Wallace et al., 2016).

CELLULOSE DEGREE OF POLYMERIZATION

The degree of polymerization (DP) is the number of glucose units in a cellulose molecule. Cellulose is a linear homopolymer of anhydroglucose units linked by β-1,4 linkages with a DP of up to 20,000. The DP of cellulose depends on the biomass type e.g., hardwood, softwood or grass. Regardless of the natural variation in the DP, the method for DP determination requires cellulose to be isolated, which results in chain splitting of the cellulose macromolecules. Therefore, the DP of cellulose is dependent of the isolation method applied. Cellulose is also composed of polydisperse molecules with varying molecular masses, and as such the DP is not homogeneous and can vary between native cellulose and cellulose originating from processed biomass. The average of the molecular mass is calculated as a function of the DP and the molecular mass of the anhydroglucose. For this reason, the molecular mass is expressed as an average of the various molecular chain sizes.

The DP of cellulose is an important parameter in the determination of the effect of pretreatment on the biomass, biomass enzymatic digestibility and the properties of cellulose composite materials (Wallace et al., 2016; Brienzo et al., 2015, Shinoda et al., 2012; Hallac and Ragauskas, 2011). The cellulose DP has also been reported to affect the mechanical properties of the nanofibrillated materials (Henriksson et al., 2008), as well as the enzymatic hydrolysis process (Hallac and Ragauskas, 2011; Puri, 1984). Suggestion have also been made that it influences the cellulose accessibility since the crystalline regions could be difficult for enzymes to access. Therefore, pretreatment is desired to fractionate the material, remove or modify hemicelluloses and lignin, and thereby exposing the cellulose to enzyme action. While some pretreatment procedures, such as dilute acid and steam explosion produce some of these desired effects on the biomass, they also result in a decrease in the cellulose DP (Brienzo et al., 2015; Wallace et al., 2016). Pretreatment procedures such as steam explosion, alkaline, carbon dioxide, and ozone were reported to decrease the DP of cellulose. However, pretreated materials with lower DPs have been reported to result in a better enzymatic hydrolysis rate (Puri, 1984). Alkaline pretreatment has been reported to cause cellulose DP decrease through two reactions that contribute to cellulose chain scission (breaking β-1,4 linkage) generating reducing end and decrease DP (Testova et al., 2014). The primary peeling is responsible for reduction in chain length through a progressive removal of glucose units, and yet still forming reducing end. This process is stopped through the stopping reaction. Secondary peeling is responsible for breaking down of internal β-1,4 linkages and the random formation of new end group that are susceptible to primary peeling. Primary peeling is therefore responsible for the yield loss while secondary peeling is responsible for DP decreases.

The cellulose DP can affect the enzymatic hydrolysis primarily because of the reactivity. A lower cellulose DP has a higher number of reducing ends resulting in an increase in the available substrate sites for exoglucanases action. Similarly, endoglucanase breaks the internal β-1,4 linkages, reducing the DP, creating reducing ends for the exoglucanase (Pala et al., 2007). The reactivity of cellulose could also be influenced by the inter and intra-molecular interaction between cellulose chains. Short cellulose chains are supposed to form week interactions than long chains, thereby providing higher accessibility or susceptibility of the cellulose to enzyme action (Puri, 1984). These factors contribute to a lower DP cellulose showing a higher yield and rate of enzymatic hydrolysis than higher DP cellulose chains (Pan et al., 2008). Pretreatment results in cellulose DP reduction and the creation of more reducing ends groups. Pretreatment processes such as dilute acid, SO_2 and steam explosion are able to break the β1,4 linkages, generating more reducing ends, compared to pretreatments with high pH (alkaline based) (Kumar et al., 2009).

Some works suggest that cellulose DP is less important to enzymatic hydrolysis than other physicochemical properties (Mansfield et al., 1999; Brienzo et al., 2015; Wallace et al., 2016). The pretreatment changes the cellulose DP, however, other physicochemical properties are also modified and all together influence the material digestibility. Hemicelluloses and lignin removal/modification probably are the most important chemical modification that increases the cellulose accessibility (Brienzo et al., 2015). These chemical and physicochemical properties can musk the DP effect, which is difficult to access in terms of its relative effect on the digestibility. A similar challenge arises with regards to the establishment of the effect of the DP during the enzymatic hydrolysis. The DP generally decreases during the enzymatic hydrolysis. Consequently, the rate and the increment in the glucose yield decrease as it progresses. During enzymatic hydrolysis, the cellulose is depolymerized, resulting in a decrease in its overall content. On the other hand, the lignin enriches in the unhydrolyzed material, decreasing the cellulose accessibility (Wallace et al., 2016). These factors are more important on comparison to the cellulose DP effect. Therefore, the evaluation of its effect on the enzymatic hydrolysis kinetics can be ignored.

Cellulose Accessibility

The use of lignocellulosic materials for liquid biofuels production or macromolecule extraction to produce high value chemicals needs an intermediate step such as pretreatment to improve the overall conversion process. The lignocellulosic material is naturally resistant to the bioconversion process due its high recalcitrance (Sant'Anna et al., 2014; Zheng et al., 2009). Bioprocessing, involving enzymatic or microbial action, can be successful with a pretreatment step that disrupts the plant cell wall structural assembly thereby exposing the polysaccharides. Pretreatment contributes to the deconstruction of the plant cell wall and thereby exposing cellulose. Cellulose accessibility has been suggested to be an important property of the lignocellulosic material that determines the enzymatic hydrolysis success action (Chandra et al., 2008; Brienzo et al., 2015). Pretreatment can alter the cellulose chemical and physical composition or the overall physical structure of the biomass. The accessibility property is influenced by several characteristic of the material and the physical nature of the cellulose. The surface area of the cellulose influences the accessibility and it can be measured as internal or external surface area (Chandra et al., 2008). Pores present in the material play important hole on the accessibility. Pores can be measured based on size and volume distribution. Moreover, material deconstruction as identified through the pores structure and visual danification can be correlated to accessibility based on dye adsorption and enzymatic hydrolysis glucose yield (Figure 1).

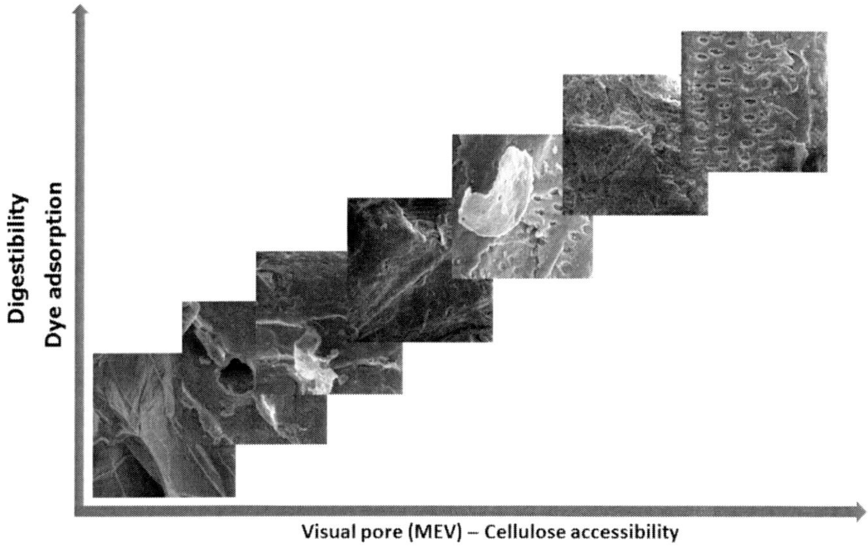

Figure 1. Lignocellulosic material digestibility direct correlation to accessibility (dye adsorption) and visual deconstruction.

The direct contact between cellulase and cellulose is necessary to enable the catalysis and break the β-1,4 linkage. However, the substrate is in a solid phase and enzyme is in a liquid phase. Cellulase needs physical contact with cellulose before a reaction can take place. In the catalytic reaction it is necessary that the enzyme adsorb onto the cellulose surface and break the monomer linkages and thereby releasing oligomers and cellobiose, and finally desorption to the liquid phase (Sun and Cheng, 2002). In this complex mechanism, the first part in relation to enzyme desorption on cellulose surface, is critical for the reaction occur. The effectiveness of the catalyst is, according to that is directly dependent of the cellulose exposure, or accessibility through its specific available surface area. Several works observed a direct correlation between the enzymatic hydrolysis glucose yield and the initial pore volume, interior specific surface area, exterior specific surface area, total specific surface area (internal + external) and the ratio of internal/external specific surface area (Brienzo et al., 2015; Zhao et al., 2012; Chandra et al., 2008; Grethlein, 1985).

The specific surface area of the cellulose is divided into the internal specific and the external specific surface areas. The internal specific surface area is related to the capillary structure of the cellulose and pores that are enhanced by the pretreatment process. The external specific surface area is related to the exposure of the cellulose fibre. The exposure of the cellulose fibres is dependent on the nature of chemical and physical barriers such as the amount of lignin and hemicelluloses present in the biomass. Biomass physical properties such as particle size also influence the exposure of cellulose since the reduction of the particle size increases the contact surface. The total surface area i.e., the sum of the internal and external specific surface areas, can be used as great tool to evaluate biomass response to pretreatment (Brienzo et al, 2015 and 2014, Chandra et al., 2008). The internal and external surface areas can undergo modification based on the type of pretreatment process employed and the selected severity of the conditions.

Dilute acid pretreatment promotes hemicelluloses removal and lignin reallocation and thereby increasing the external and internal surface areas (Brienzo et al., 2015). Pretreatment with polar solvents causes fibre swelling and a probably increase the internal specific surface area. A comparison of different processes indicated that for example, acid pretreatment is more effective than hot water and alkaline pretreatments in increasing cellulose accessibility (Meng et al., 2015). Some works suggested that hemicelluloses removal is more important than lignin (Meng et al., 2015; Leu and Zhu, 2013). It is important to note that acid pretreatment provokes lignin reallocation while hemicelluloses is removed. Alkaline pretreatment can solubilize both the hemicelluloses and lignin, and hot water is more effective in hemicelluloses removal.

The internal and external specific surface areas can change also as influenced by enzyme hydrolysis. The enzymatic hydrolysis kinetics can be divided into 3 phases i.e., a fast initial hydrolysis rate, a moderate hydrolysis rate and a slow or recalcitrant phase (Wallace et al., 2016). Steam exploded sugarcane bagasse was characterized as rich in internal specific surface area, until the intermediate phase is reached, where after it was characterized as predominantly consisting of external specific surface area in studies by Wallace et al., (2016). The authors explained the slowdown of enzymatic hydrolysis of steam exploded sugarcane bagasse as a consequence of lignin enrichment, which consequently decreased the accessibility of the residual cellulose that remaining unhydrolyzed. The slowdown in enzymatic hydrolysis is also attributed to enzyme deactivation. In fact, there is no single factor that could explain the slowdown of the enzymatic hydrolysis process, but a combination of them (Wallace et al., 2016). The factors involved in the enzymatic slowdown in hydrolysis are related to the change in the environment under which the enzyme operates. These can cause enzyme deactivation (thermal and mechanical deactivation), unspecific bonding to lignin, end-product inhibition. Further, the synergism between cellulase enzymes and substrate as related to pore volume and distribution, fiber organization, crystallinity, particle size, internal specific surface area, external specific surface area also have an effect. Some of these substrate related factors have an indirect or direct influence on cellulose accessibility.

The accessibility could be more specific considering specific surface area and accessible surface area, which are related to cellulose accessible area. The term "cellulose accessible area" should be applied when methods that take in account the cellulose surface accessible to cellulase are used. The substrate accessibility is not limited to cellulose surface accessible to cellulase, meaning that substrate accessibility includes surfaces that is not accessible to cellulase. The pore diameter also has an influence on cellulase enzyme access to the cellulose. Considering that the diameter average of cellulase enzymes from fungi is 5.1 nm, the pore size diameter is a limiting factor for enzymatic hydrolysis (Grethlein, 1985; Chandra et al., 2008). A larger diameter will be an advantage to cellulose conversion to glucose. The substrate accessibility can be directly measured based on the accessible pore volume (porosity) or surface area of the substrate using a set of probing molecules (water retention value, differential scanning calorimetry and solute exclusion technique). Indirect measurement which are based on the determination of the adsorption of a molecule to the material (nitrogen adsorption BET, and dyes adsorption Simon's staining) can also be applied (Leu and Zhu, 2013). To determine the cellulose accessible area, it is necessary to exclude the non-cellulosic area, and the probe molecule needs a cellulose specific binding module, and diameter of 5.1 nm to simulate the enzyme diameter (Hong and Zhang, 2007). Considering possible to avoid the unspecific bind, cellulase can be used to quantify the accessible surface area (Wang et al., 2012a).

Lignin Content, Composition and Lignin-Carbohydrate-Complex Impact on Enzymatic Hydrolysis

The lignin content, its composition as well as the presence and the composition of lignin carbohydrate complexes associated with residual lignin in lignocellulose from wood and groups of herbaceous plants, contribute to the recalcitrance of lignocellulose, and consequently efficient conversion of cellulose. Earlier studies pinned the lignin contribution to recalcitrance mainly to its content and contended that the structure had little or no contribution. Studies on forage digestibility established negative relationships between lignin content and rumen digestibility, but no relationship between lignin composition and digestibility (Reddy et al., 2005). Chen and Dixon (2007) also found that down regulating lignin production in biomass improves fermentable sugar yields for biofuel production. However, they argued that only lignin content had a major influence. Nakagame et al., (2010) investigated the influence of the residual lignin remaining in pretreated lignocellulosic substrates of corn stover, poplar, and lodgepole pine on subsequent enzymatic hydrolysis and found that in fact the nature of the residual lignin obtained after pretreatment, based on source and pretreatment method significantly influenced hydrolysis. Lignin isolated from lodgepole pine and steam pretreated poplar decreased the hydrolysis yields of Avicel, whereas the isolated lignins from corn stover did not appear to decrease the hydrolysis yields significantly. In a subsequent study using isolated lignins from Douglas fir, a similar trend was observed whereby there was a decrease in the hydrolysis of Avicel (Nakagame et al., 2011). Further, higher pretreatment severities tended to decrease the enzymatic hydrolysis of Avicel to a greater degree due to increased condensation of the lignin under such conditions, while milder conditions affected the hydrolysis of the substrate the least. Studer et al., (2011) observed a strong negative correlation between sugar release and lignin content for pretreated *Populus trichocarpa* samples with an S/G ratio < 2.0. For higher S/G ratios, sugar release was generally higher, and the negative influence of lignin was less pronounced.

In general, the efficiency of processing is directly proportional to the amount of syringyl units in lignin (Wang et al., 2012b) which is regarded as a highly reactive lignin building unit (Sun et al., 2000). This is due to the presence of the methoxyl group at C5 in the syringyl lignin which results in a low degree of condensation, and therefore the syringyl lignin is easily hydrolyzed compared to guaiacyl and p-hydroxyphenyl lignin. The guaiacyl units have a free C-5 position available for C-C inter-unit bonds, which makes them fairly resistant to lignin depolymerisation (Fengel and Wegener, 1984).

Residual lignin isolated from the lignocelluloses is assumed to be chemically bonded to carbohydrates and mostly to xylan. In general, residual lignin associated with glucose is easily hydrolyzed compared to lignin that is associated with xylan (Lawoko et al., 2011; Li et al., 2011). Chen et al., (2007) observed that enzymatic hydrolysis released more xylose from transgenic lines where lignin was down regulated than from control lines, suggesting that lignin modification increases the accessibility of residual hemicelluloses to degradative enzymes. This in turn suggests that a lower lignin content is correlated to a low the xylan-lignin LCC content in the biomass, resulting in ease in xylose release. The high resistance of the xylan lignin LCC is reported to be due to the β-O-4 bond type between lignin and xylan, which results to a linear coupling mode structure (Lawoko, 2005). Studies conducted by Santos et al., (2012) have established a correlation between the structural forms of β-O-4- structures found in lignin and S/G ratio, indicating that the syringyl/guaiacyl composition is the key factor governing the proportion of β-O-4- structure of lignin. S/G ratio is a key factor affecting the rate of biomass processing, be it pulping, enzymatic hydrolysis or pre-treatment (Santos et al., 2011). It also seems that the S/G ratio appears to control the β-O-4- content and the degree of condensation (Santos et al., 2012).

In light of the foregoing discussion, Li et al., (2010) suggested that increasing S monomer content in lignin through genetic engineering may perhaps be a promising approach to improve enzymatic hydrolysis efficiency, which in turn should reduce the cost of biomass conversion to biofuels. Palonen (2004) had earlier proposed the use of oxidative lignin modifying enzymes and partial removal of the lignin using laccase mediator treatments. It is also clear that the severity of the biomass pretreatment conditions also has an effect on the structure of the lignin, which in turn affects enzymatic hydrolysis. The value prior puling approach, which involves the extraction of the hemicelluloses for further beneficiation, prior to further processing of the biomass such as in pulping or enzymatic hydrolysis, seems to hold promise in increased efficiency due to the reduced xylan linked LCC and partial reduction of lignin, and higher lignin S/G ratios in the residual biomass (Njamela 2015).

Conclusion

Cellulose is the main structural carbohydrate component of plant cell walls. Cellulose functions in the plant is related to protection of the cells by formation of structures of the cell wall. The morphology of the plant is influenced by the cells organization and due the cell walls shapes, characterized by layers in different cellulose fiber organization. The cellulose layers have different content of cellulose and fiber orientation that together promote higher mechanical stability to the walls. The cellulose is highly organized in the plant cell walls with components, such as pectin, hemicellulose, proteins and lignin. The amount, type and variation of these components impose physical barrier protecting cellulose from microorganisms and its enzymes. Cellulose molecules have its intrinsic recalcitrance based on a highly organized structure (crystalline) compared with less proportion of amorphous region, as well as the degree of polymerization. Moreover, the degree of recalcitrance is also influenced by the association with other carbohydrates (hemicelluloses) and lignin. The heterogeneity of the biomass even for the same specie (tissue, type of cells) adds complexity to the lignocellulose biomass.

The cellulose in lignocellulosic material structure is protected by the hemicelluloses and lignin macromolecules, decreasing its accessibility to enzyme action. A pretreatment step is necessary to increase the cellulose accessibility, modify the fibers organization, remove partially or completely the hemicellulose shield and lignin barrier. Cellulose enzymatic hydrolysis is a challenge even for pure cellulose due its molecule organization and DP. The enzymes mechanism and tools for cellulose hydrolysis relies not only on cellulases, but also so called auxiliaries enzymes and proteins. The cellulase industrial application face difficulties related to its performance, which is dependent on the substrate characteristics and properties (pure cellulose or lignocellulosic material), and the cost of final product, which impact in the economic feasibility of the enzyme production and application.

References

Alvira, P., Tomás-Pejó, M., Ballesteros, M. & Negro, M. J. (2010). Pretreatment technologies for an efficient bioethanol production process based on enzymatic hydrolysis: A review. *Bioresour Technol*, 101, 4851-4861.

Atalla, R. H. & Vander Hart, D. L., (1984). Native cellulose: A composite of two distinct crystalline forms. *Science,* 223, 283-285.

Benjamin, Y., Cheng, H. & Görgens, J. F. (2013). Evaluation of bagasse from different varieties of sugarcane by dilute acid pretreatment and enzymatic hydrolysis. *Ind Crop Products.*, 51, 7-18.

Berlin, A., Gilkes, N., Kilburn, D., Maximenko, D., Bura, R., Markov, A., Skomarovsky, A., Gusakov, A. Sinitsyn, A., Okunev, O., Solivieva, I., & Saddler, J. N. (2006). Evaluation of cellulase preparations for hydrolysis of hardwood substrates. *Appl Biochem Biotechnol*, 129-132, 528-545.

Brienzo, M., Tyhoda, L., Benjamin, Y. & Görgens, J. (2015). Relationship between physicochemical properties and enzymatic hydrolysis of sugarcane bagasse varieties for bioethanol production. *N Biotechnol., 32*, 253-262.

Brienzo, M., Abud, Y., Ferreira, S., Corrales, R. C. N. R., Ferreira-Leitão, V. S., De Souza, W. & Sant'Anna, C. (2016). Characterization of anatomy, lignin distribution, and response to pretreatments of sugarcane culm node and internode. *Ind Crop products*. 84, 305-313.

Chandra, R., Ewanick, S., Hsieh, C. & Saddler, J. N. (2008). The characterization of pretreated lignocellulosic substrates prior to enzymatic hydrolysis. Part 1. A modified Simons' staining technique. *Biotechnol Progr*, 2008;24:1178–85.

Chang, V, S, & Holtzapple, M. T. (2000). Fundamental factors affecting biomass enzymatic reactivity. *Appl Biochem Biotechnol.*, 84-86, 5-37.

Chen, F. & Dixon, R. A. (2007). Lignin modification improves fermentable sugar yields for biofuel production. *Nat Biotechnol.* 25, 759-761.

Chundawat, S. P. S., Bellesia, G., Uppugundla, N., Sousa, L., Gao, D. & Cheh, A. (2011). Restructuring the crystalline cellulose hydrogen bond network enhances its depolymerization rate. *J Am Chem Soc*, 133, 11163-11174.

Costa, T. H. F., Masarin, F., Bonifácio, T. O., Milagres, A. M. F. & Ferraz, A. (2013). The enzymatic recalcitrance of internodes of sugar cane hybrids with contrasting lignin contents. *Ind Crop Prod.* 51, 202-211.

De Souza, W. & Sant'Anna, C. (2012). Microscopy as a tool to follow deconstruction of lignocellulosic biomass. In: Current microscopy contributions to advances in science and technology. Antonio Méndez-Vilas (ed), Espanha, Formatex Research Center, 639-645.

Fan, L. T., Lee, Y. H. & Beardmore, D. H. (1980). Mechanism of the enzymatic hydrolysis of cellulose – effects of major structural features of cellulose on enzymatic hydrolysis. *Biotechnol Bioeng*., 22, 177-199.

Fan, L. T., Lee, Y.-H. & Beardmore, D. R., (1981). The influence of major structural features of cellulose on rate of enzymatic hydrolysis. *Biotechnol Bioeng*, 23, 419-424.

Fengel, D. & Wegener, G. (1984). Wood: chemistry, ultrastructure, reactions. Berlin, New York: Walter de Gruyter.

Fierobe, H. P., Bayer, E. A., Tardif, C., Czjzek, M., Mechaly, A., Bélaïch, A., Lamed, R., Shoham, Y. & Bélaïch, J. P. (2002). Degradation of cellulose substrates by cellulosome chimeras. Substrate targeting versus proximity of enzyme components. *J Biol Chem*, 277, 49621-49630.

Grethlein, H. E. (1985). The effect of pore size distribution on the rate of enzymatic hydrolysis of cellulosic substrates. *Nat Biotechnol*, 3, 155-160.

Guo, F., Shi, W., Sun, W., Li, X., Wang, F., Zhao, J, & Qu, Y. (2014). Differences in the adsorption of enzymes onto lignins from diverse types of lignocellulosic biomass and the underlying mechanism. *Biotechnol Biofuels*. 7: 38.

Hall, M., Bansal, P., Lee, J. H., Realff, M. J., Bommarius, A. S. (2010). Cellulose crystallinity – a key predictor of the enzymatic hydrolysis rate. *FEBS J*, 277, 1571-82.

Hallac, B. B. & Ragauskas, A. J. (2011). Analyzing cellulose degree of polymerization and its relevancy to cellulosic ethanol. *Biofuels Bioprod Biorefin*, 5, 215-25.

Henriksson, M., Berglund, L. A., Isaksson, P., Lindström, T., Nishino, T. (2008). Cellulose nanopaper structures of high toughness. *Biomacromolecules*, 9, 1579-1585.

Himmel, E. M Shi-You, D, Johnson, D. K., Adney, W. S., Nimlos, M. R., Brady, J. W. & Foust T. D. (2007). Biomass recalcitrance: Engineering plants and enzymes for biofuels production, *Science,* 315, 804-807.

Hong, J., Ye, X. & Zhang, Y-H. P. (2007). Quantitative determination of cellulose accessibility to cellulase based on adsorption of a nonhydrolytic fusion protein containing CBM and GFP with its applications. *Langmuir*, 23, 12535-12540.

Howard, R. L., Abotsi E., van Rensburg, J. E. L. & Howard S. (2003). Lignocellulose biotechnology: issues of bioconversion and enzyme production. *Afr J Biotechnol*, 2, 602-619.

Kumar, R., Mago, G., Balan, V. & Wymand, C. E. (2009). Physical and chemical characterizations of corn stover and poplar solids resulting from leading pre-treatment technologies. *Bioresour Technol*, 100, 3948-3962.

Langan, P., Gnanakaran, S., Rector, K. D., Pawley, N., Fox, D. T., Cho, D. W., & Hammel, K. H. (2011). Exploring new strategies for cellulosic biofuels production. *Energy and Environmental Science*, 4, 3820-3833.

Laureano-Perez, L., Teymouri, F., Alizadeh, H. & Dale, B. E. (2005). Understanding factors that limit enzymatic hydrolysis of biomass: characterization of pretreated corn stover. *Appl. Biochem. Biotechnol.*, 121-124, 1081-1099.

Lawoko, M. (2005). Lignin polysaccharide networks in softwoods and chemical pulps, characterisation, structure and reactivity. PhD Thesis, Royal Institute of Technology (KTH), Stockholm.

Lawoko, M. & van Heiningen, A. R. P. (2011). Fractionation and characterization of completely dissolved ball milled hardwood. *J Wood Chem Technol,* 31, 183-203.

Leu, S-Y. & Zhu, J. Y. (2013). Substrate-related factors affecting enzymatic saccharification of lignocelluloses: Our recent understanding. *Bioenerg. Res.*, 6, 405-415.

Li, J., Martin-Sampedro, R., Pedrazzi, C. & Gellerstedt, G. (2011). Fractionation and characterisation of lignin-carbohydrate complexes (LCCs) from eucalyptus fibres. *Holzforschung,* 65, 43-50.

Lynd, L. R., Laser, M. S., Bransby, D., Dale, B. E., Davison, B. & Hamilton, R. (2008). How biotech can transform biofuels. *Nat Biotechnol.*, 26, 169-172.

Nakagame, S., Chandra R. P. and Saddler, J. N. (2010). The effect of isolated lignins, obtained from a range of pretreated lignocellulosic substrates, on enzymatic hydrolysis. *Biotechnol Bioeng,* 105, 871-879.

Nakagame, S., Chandra R. P., Kadla, J. F. & Saddler, J. N. (2011). Enhancing the enzymatic hydrolysis of lignocellulosic biomass by increasing the carboxylic acid content of the associated lignin. *Biotechnol Bioeng*, 108, 538–548.

Njamela, S. N. (2015). Lignin polysaccharide networks in biomass and corresponding processed materials. Department of Forest and Wood Science. Stellenbosch, South Africa, University of Stellenbosch. MSc Thesis.

Masarin, F., Gurpilhares, D. B., Baffa, D. C. F., Barbosa, M. H. P., Carvalho, W., Ferraz, A. & Milagres, A. M. F. (2011). Chemical composition and enzymatic digestibility of sugarcane clones selected for varied lignin content. *Biotechnol Biofuels*, 4, 55.

Meng, X., & Ragauskas, A. J. (2015). Recent advances in understanding the role of cellulose accessibility in enzymatic hydrolysis of lignocellulosic substrates. *Cur Opin Biotechnol,* 27, 150-158.

Min, D.-Y., Li, Q., Chiang, V., Jameel, H., Chang, H.-M., & Lucia, L. (2014). The influence of lignin-carbohydrate complexes on the cellulase-mediated saccharification I: Transgenic black cottonwood (western balsam poplar, California poplar) *P. trichocarpa* including the xylan down-regulated and the lignin down-regulated lines. *Fuel*, 119, 207-213.

Mooney, C. A., Mansfield, S. D., Touhy, M. G. & Saddler, J. N. (1998). The effect of initial pore volume and lignin content on the enzymatic hydrolysis of softwoods. *Bioresour Technol.,* 64, 113−119.

Pala, H., Mota, M. & Gama, F. M. (2007). Enzymatic depolymerisation of cellulose. *Carbohyd. Polym*., 68, 101-108.

Palonen, H., Tjerneld, F., Zacchi, G., Tenkanen, M. (2004). Adsorption of *Trichoderma reesei* CBH I and EG II and their catalytic domains on steam pretreated softwood and isolated lignin. *J Biotechnol,* 107, 65-72.

Pan, X., Xie, D., Yu, R. W. & Saddler, J. N. (2008). The bioconversion of mountain pine beetle-killed lodgepole pine to fuel ethanol using the organosolv process. *Biotechnol Bioeng*., 101, 39-48.

Perez, S. & Mazeau, K. (2005). Conformations, structures, and morphologies of celluloses. In: Polysaccharides: Structural diversity and functional versatility. Edited by: S. Dumitriu. Marcel Dekker, New York.

Perez, S. & Mazeau, K. (2004). Conformation, structure and morphologies of cellulose. In: Polysaccharides: Structural Diversity and Functional Versatility, Dimitriu (ed), Macel Dekker, 41-68.

Puri, V. P. (1984). Effect of crystallinity and degree of polymerization of cellulose on enzymatic saccharification. *Biotechnol Bioeng*, 26, 1219-1222.

Reddy, M. S. S., Chen, F., Shadle, G., Jackson, L., Aljoe, H. & Dixon, R. A. (2005). Targeted down-regulation of cytochrome P450 enzymes for forage quality improvement in alfalfa (*Medicago sativa* L.) *Proc. Natl Acad Sci,* 102, 16573–16578.

Sant'Anna, C., De Souza, W. & Brienzo, M. (2014). The Influence of the Heterogeneity, Physicochemical and Structural Properties on the Recalcitrance and Conversion of Sugarcane Bagasse In: Sugarcane: Production, Consumption and Agricultural Management Systems.1 ed.: Nova Science Publishers, 1-33.

Shinoda, R., Saito, T., Okita, Y. & Isogai, A. (2012). Relationship between length and degree of polymerization of tempo-oxidized cellulose nanofibrils. *Biomacromolecules*, 13, 842-849.

Sjostrom, E. (1993). Wood Chemistry. Fundamentals and Applications. Second edition ed. 1993, San Diego: Academic press. 292.

Santos, R. B., Capanema, E. A., Balakshin, M. Y, Chang, H.-M. and Jameel, H. (2012). Lignin structural variation in hardwood species. *Journal of Agricultural and Food Chemistry* 60(19). 4923-2930.

Studer, M. H., DeMartini, J. D., Davis, M. F., Sykes, R. W., Davison, B., Keller, M., Tuskan, G. A., Wyman, C. E. (2011) Lignin content in natural Populus variants affects sugar release. *Proc Natl Acad Sci* USA, 108, 6300-6305.

Sun, R.-C., Tomkison, J., Sun, X. F. and Wang, N. J. (2000). Fractional isolation and physico-chemical characterisation of alkali-soluble lignins from fast-growing poplar wood. *Polymer,* 41, 8409-8417.

Sun, Y. & Cheng, J. (2002). Hydrolysis of lignocellulosic materials for ethanol production: A review. *Bioresour Technol*, 83, 1-11.

Testova, L., Nieminen, K., Penttila, P. A., Serimaa, R., Potthast, A. & Sixta, H. (2014). Cellulose degradation in alkaline media upon acidic pretreatment and stabilisation. *Carbohyd Polym*, 100, 185-194.

Ververis, C., Georghiou, K., Christodoulakis, N., Santas, P. & Santas, R. (2004). Fiber dimensions, lignin and cellulose content of various plant materials and their suitability for paper production. *Ind Crop Prod.*, 19, 245-254.

Waldron, K., (2010). Bioalcohol production: Biochemical conversion of lignocellulosic biomass. North America, 6000 Broken Sound Parkway, NW, Suite 300, Boca Raton, FL 33487, USA: Woodhead Publishing Limite.

Wallace, J., Brienzo, M., Garcıa-Aparicio, M. & Görgens, J. (2016). Lignin enrichment and enzyme deactivation as the root cause of enzymatic hydrolysis slowdown of steam pretreated sugarcane bagasse. *New Biotechnol*. 33, 2016.

Wang, Q. Q., He, Z., Zhu, Z., Zhang, Y-H. P., Ni, Y. & Luo, X. L. (2012a). Evaluations of cellulose accessibilities of lignocelluloses by solute exclusion and protein adsorption techniques. *Biotechnol Bioeng*., 109, 381-389.

Wang, K., Yang, H. Y., Yao, X., Sun, R. C. & Jones, G. L. (2012b). Structural characterisation of isolated lignin from *Caragana korshinskii kom. Cell Chem Technol,* 46, 185-191.

Wang, S., Lee, S-H. & Cheng Q. (2010). Mechanical properties of cellulosic materials at micro- and nanoscale levels. In: Cellulose, structure and properties. Lejeune, A., Deprez, T. Nova Science Publishers.

Ward, K. J. R. (1950). Crystallinity of cellulose and its significance for the fiber properties. *Tex Res J*, 20, 363-372.

Wu, L., Arakane, M., Ike, M., Wada, M., Takai, T., Gau, M. & Tokuyasu, K. (2011). Low temperature alkali pretreatment for improving enzymatic digestibility of sweet sorghum bagasse for ethanol production. *Bioresour Technol*, 102, 4793-4799.

Yoshida, M., Liu, Y., Uchida, S., Kawarada, K., Ukagami, Y., Ichinose, H., Kaneko, S. & Fukuda, K. (2008). Effects of cellulose crystallinity, hemicellulose, and lignin on the enzymatic hydrolysis of *Miscanthus sinensis* to monosaccharides. *Biosci. Biotechnol. Biochem.*, 72, 805-810.

Zeng, M., Ximenes, E., Ladisch, M. R., Mosier, N. S., Vermerris, W., Huang, C-P. & Sherman, D. M. (2012). Tissue-specific biomass recalcitrance in corn stover pretreated with liquid hot-water: Enzymatic hydrolysis (Part 1). *Biotechnol Bioeng.*, 109, 390-397.

Zhang, Y. H. P. & Lynd, L. R. (2004). Toward an aggregated understanding of enzymatic hydrolysis of cellulose: noncomplexed cellulase systems. *Biotechnol Bioeng.*, 88, 797–824.

Zhan, H. Y. (2005). Fiber chemistry and physics. Beijing: Science Press.

Zhao, X., Zhang, L. & Liu, D. (2012). Biomass recalcitrance. Part I: the chemical compositions and physical structures affecting the enzymatic hydrolysis of lignocellulose. *Biofuels Bioprod Bioref*, 6, 465-482.

Zheng, Y., Pan, Z. & Zhang, R. (2009). Overview of biomass pretreatment for cellulosicethanol production. *Int J Agric Biol Eng*, 2, 51-68.

RELATED NOVA PUBLICATIONS

CELLULASE: TYPES AND ACTION, MECHANISM, AND USES

Adam E. Golan

Cellulase refers to a class of enzymes produced chiefly by fungi, bacteria, and protozoans that catalyze the cellulolysis (or hydrolysis) of cellulose. The enormous potential that cellulases have in biotechnology is the driving force for continuous basic and applied research on these biocatalysts from fungi and bacteria. Cellulases are found in many fields, such as animal feeding, brewery and wine, food, textile and laundry, pulp and paper products. The growing interest toward the conversion of lignocellulosic biomass into fermentable sugars has generated an additional request for cellulases and their related enzymes. This book presents research in the study of cellulase, including biotechnological applications of microbial cellulases; using agro-industrial by-products as raw material for cellulase production; and the enzyme saccharification of cereal crop residues using dilute alkali pretreatment. (Imprint: Nova)

Cellulose and Cellulose Derivatives: Synthesis, Modification and Applications

Md. Ibrahim H. Mondal
Professor, Department of Applied Chemistry and Chemical Engineering, Rajshahi University, Rajshahi, Bangladesh

Cellulose is the principal constituent of all plant life; it is the most abundant, important and fascinating biopolymer on earth. Cellulose, as an almost inexhaustible, environmentally benign and renewable material, has stimulated basic and applied research as well as inspired significant progress in Polymer Science. In recent years, cellulose has gained renewed importance as a raw material. Although ground breaking research is carried out on cellulose, it still possesses high potential for future applications; it can be easily modified to more natural and sustainable alternatives compared to synthetic products by certain techniques. The present book reviews some vital issues and topics on the latest science and technological advances in cellulose and its derivatives. This catalog acts as an essential source of information to readers in the exploration for possible applications of cellulose and its derivatives. The authors hope this collection will spark a generation of new ideas for product development.

The present book contains 25 invited contributions written by leading experts in the field of cellulose and cellulose derivatives. It is divided into three parts: Part I, Cellulose Synthesis and Modification; Part II, Cellulose Derivatives; and Part III, Applications of Cellulose Derivatives. Highlights of this book include the mechanism of cellulose formation in biosynthetical processes; surface modification and functionalization of cellulose fibers; advances in the homogenous and heterogeneous phase modification of cellulose to create unusual and functional derivatives; analysis and characterization of modified derivatives; derivatives for antimicrobial, medical and pharmaceutical applications, and wastewater treatment; dendronized and hyperbranched cellulose derivatives; and rheology of nanocellulosic systems.

The present book covers a wide range of material, and hence a great number of people — both academics and industrialists — may find topics of interest on cellulose and cellulose derivatives within. (Imprint: Nova)

INDEX

A

accessibility, 78, 88, 89, 97, 98, 100, 102, 105, 107, 108, 109, 110, 111, 113, 114, 116, 118
acclimatization, 55
acetic acid, 43
acetylation, 98
Achromobacter, viii, 2, 21, 22
acid, 4, 8, 10, 12, 23, 25, 26, 43, 61, 67, 74, 75, 78, 102, 105, 107, 110, 115, 117
acidic, 14, 48, 67, 119
Acidothermus, viii, 2, 14, 40
Acinetobacter, viii, 2, 15, 16, 42, 59
Acremonium, viii, 2, 17, 18
adsorption, 86, 98, 100, 105, 109, 111, 116, 119
advancements, 84
Aeromonas, viii, 2, 14, 16, 37
agar, vii, 1, 3, 4, 5, 6, 7, 8, 9, 10, 11, 12, 13, 15, 37, 48
alfalfa, 118
algae, 4, 67, 104
algorithm, 42
alkaline media, 119
Alternaria, 17, 19, 25, 26, 42
amino, 74, 75
ammonium, 21, 22, 23, 24, 26, 53
amylase, 18, 42, 61
anaerobic bacteria, 47
anaerobic digesters, 60
anaerobic digestion, 47
anatomy, 101, 115
Aneurinibacillus, viii, 2, 14, 16, 32
Anoxybacillus, viii, 2, 14, 16, 21, 22
antibiotic, 57, 59
Aspergillus, viii, 2, 17, 18, 20, 24, 25, 26, 28, 29, 32, 34, 37, 39, 40, 41, 43, 47, 48, 52, 59, 72, 73
Aspergillus terreus, 25, 26, 40, 43

B

Bacillus, viii, 2, 15, 16, 17, 18, 20, 21, 22, 28, 32, 33, 34, 36, 37, 38, 39, 42, 43, 47, 52, 57, 58, 59, 60, 61, 62, 67, 72
Bacillus subtilis, 15, 17, 21, 33, 57, 58, 59, 61
background information, 3
bacteria, viii, 2, 3, 4, 5, 6, 7, 14, 15, 20, 26, 30, 31, 32, 33, 34, 37, 38, 39, 40, 41, 42, 43, 44, 45, 46, 47, 48, 50, 52, 53, 56, 57, 58, 59, 60, 61, 72, 73, 76, 85, 104, 121
bacterial cells, 50, 51
bacterial colonies, 6
bacterial strains, 2, 58, 62
bacterium, 14, 22, 23, 29, 31, 32, 42, 44
Bacteroides, 31, 76
Bangladesh, 17, 36, 122

Bifidobacterium, 31
bioassay, 53
bioavailability, 73
biocatalysts, 26, 64, 121
biochemistry, 32, 88
bioconversion, vii, 1, 3, 27, 99, 108, 116, 118
biodegradability, 62
biodegradation, 54, 55, 58
biofuel, 27, 47, 69, 80, 90, 112, 115
bioinformatics, 74
biological processes, 64
biomass, viii, ix, 2, 3, 14, 27, 34, 38, 39, 41, 51, 63, 66, 67, 68, 69, 75, 79, 81, 83, 85, 86, 88, 89, 90, 91, 92, 94, 96, 98, 99, 100, 101, 102, 103, 104, 105, 106, 108, 110, 112, 113, 114, 115, 116, 117, 119, 120, 121
biomaterials, 67, 96
biopolymer, vii, 1, 122
biorefineries, 96
bioremediation, 64
biosphere, 3
biosynthesis, 43
biotechnological applications, 121
biotechnology, 3, 32, 33, 47, 55, 57, 64, 73, 88, 92, 116, 121
bonds, 2, 49, 73, 75, 77, 103, 104, 113
Brucella, viii, 2, 16
Burkholderia, 31, 47, 59

C

C. thermocellum, 79
calorimetry, 111
Candida, viii, 2, 19
capillary, 110
carbohydrate, ix, 2, 27, 53, 63, 73, 74, 75, 78, 81, 82, 84, 86, 89, 94, 98, 112, 113, 114, 117, 118
carbon, 2, 7, 13, 14, 15, 18, 19, 20, 22, 23, 24, 26, 27, 35, 39, 42, 50, 53, 55, 59, 62, 72, 79, 84, 91, 103, 107
carbon dioxide, 59, 107
carboxylic acid, 117
carboxymethyl cellulose, 6, 10, 12, 13, 19, 22, 26, 85
cardiovascular diseases, 31
casein, 7, 8, 9, 10
catabolism, 54
catalysis, 47, 67, 94, 109
catalyst, 67, 80, 109
catalytic activity, 78
cell wall, 30, 31, 66, 71, 72, 96, 98, 100, 101, 102, 105, 108, 114
Cellulase(s), v, vii, viii, ix, 1, 2, 3, 4, 5, 7, 8, 9, 10, 11, 12, 13, 14, 15, 16, 17, 18, 19, 20, 21, 22, 23, 24, 25, 26, 27, 28, 29, 30, 31, 32, 33, 34, 35, 36, 37, 38, 39, 40, 41, 42, 43, 44, 45, 46, 47, 48, 49, 50, 52, 53, 54, 55, 56, 58, 59, 60, 61, 62, 63, 65, 66, 70, 71, 72, 73, 75, 76, 78, 79, 81, 82, 83, 84, 85, 86, 88, 90, 91, 92, 93, 94, 95, 96, 99, 100, 109, 111, 114, 115, 116, 118, 120, 121
Cellulomonas, viii, 2, 5, 15, 16, 21, 22, 30, 32, 42, 47, 52, 72
cellulose, vii, viii, ix, 1, 2, 3, 4, 5, 6, 7, 8, 9, 10, 12, 13, 14, 17, 20, 22, 24, 25, 26, 27, 29, 30, 31, 33, 38, 39, 40, 41, 43, 46, 47, 48, 49, 50, 51, 52, 53, 54, 55, 56, 61, 62, 66, 67, 69, 72, 73, 75, 77, 78, 80, 81, 82, 83, 85, 86, 87, 88, 89, 90, 91, 92, 93, 94, 95, 96, 97, 98, 99, 100, 101, 102, 103, 104, 105, 106, 107, 108, 109, 110, 111, 112, 114, 115, 116, 118, 119, 120, 121, 122
Cellulose accessibility, 88, 97, 108
cellulose derivatives, 122
cellulose fibre, 100, 110
cellulose systems, 62
cellulosic biofuel, 117
Cellulosimicrobium, viii, 2, 14, 16
cellulosomes, 76, 89
CFR, 21, 22, 43
Chaetomium, viii, 2, 17, 19, 29, 41
chain scission, 107
challenges, ix, 43, 47, 53, 55, 88, 90, 91, 92, 96

chemical, x, 49, 53, 54, 66, 67, 89, 96, 97, 98, 101, 103, 108, 110, 117, 119, 120
China, 6, 16, 38, 39, 68, 94
chiral recognition, 91
chitinase, 30
chlorine, 65
chromatography, 14
climate change, 64
Clostridium, viii, 2, 14, 18, 21, 22, 31, 42, 43, 53, 58, 59, 60, 72, 73, 76
CMC, 3, 4, 5, 6, 7, 8, 9, 10, 12, 13, 15, 20, 21, 22, 23, 25, 26, 48, 53, 55
CO2, 59
coastal region, 18
commercial, vii, ix, 3, 15, 19, 27, 32, 64, 67, 68, 70, 71, 74, 82, 87
compatibility, 28, 38
competition, 54
complexity, 96, 100, 114
composition, x, 9, 23, 53, 79, 96, 98, 99, 100, 101, 108, 112, 113, 117
compost, 5, 6, 17, 32, 61, 72
compounds, 31, 47, 50, 53, 54, 68, 78, 82
compression, 102
condensation, 112, 113
configuration, 75, 87
confinement, 77
Congo, vii, 1, 3, 4, 5, 6, 7, 10, 11, 12, 13, 48, 49
construction, 59
consumption, 67, 87
contact time, 75
contamination, 13, 72
conversion rate, 72, 97
Coprococcus, 31
coproduction, 85
corn stover, 94, 102, 112, 117, 120
correlation, 98, 109, 112, 113
cost, ix, 19, 20, 63, 64, 67, 79, 80, 83, 84, 91, 97, 100, 113, 114
crystal structure, 103
crystalline, 3, 75, 78, 82, 86, 87, 92, 97, 104, 105, 106, 107, 114, 115
crystallinity, 81, 97, 98, 99, 101, 102, 105, 106, 111, 116, 118, 120
crystallinity index, 105, 106
crystals, 105
cultivation, 6, 19, 22, 42
cultural conditions, 42
culture, 5, 7, 10, 14, 16, 22, 23, 24, 28, 33, 36, 37, 48, 49, 50, 56, 90
culture conditions, 37, 90
culture medium, 22, 24, 28, 48, 50, 56
cystic fibrosis, 31
cytochrome, 118
Cytophaga, 2
cytoplasm, 49

D

deconstruction, 83, 108, 109, 115
defibrillation, 28
degradation, vii, viii, ix, 6, 13, 17, 22, 27, 30, 31, 38, 45, 47, 48, 50, 51, 52, 54, 56, 66, 67, 78, 81, 87, 88, 89, 90, 91, 96, 97, 101, 119
degradation mechanism, 90
degradation process, 54, 56
degree of polymerization, 75, 81, 97, 99, 101, 102, 106, 114, 116, 118, 119
denaturation, 83
depolymerization, vii, 1, 74, 98, 115
derivatives, 67, 122
desorption, 75, 109
detectable, 54, 56
detection, 37, 48, 60, 62
detergents, viii, ix, 3, 28, 46, 65, 66, 95
developing countries, 46
differential scanning calorimetry, 111
diffraction, 104
digestibility, viii, 2, 3, 30, 85, 88, 98, 100, 103, 105, 106, 108, 109, 112, 118, 120
digestion, 28, 47, 59, 65, 80, 105
diseases, 30, 31
distillation, 71, 96, 100
distilled water, 7, 9, 13

distribution, 14, 46, 85, 100, 101, 102, 104, 109, 111, 115, 116
diversity, vii, viii, 2, 26, 36, 87, 93, 94, 101, 118
DOI, 57
dosage, 72, 78, 80, 82, 86

E

effluents, 10, 11, 12, 42, 49, 57, 58, 60, 62, 65, 66, 79
electrophoresis, 31, 36
elongation, 104
employment, 100
energy, 30, 55, 56, 64, 65, 67, 87, 89
energy consumption, 67, 87
engineering, 55, 57, 59, 85, 86, 87, 88, 90, 113
Enhydrobacter, viii, 2, 15, 17, 21, 41
Enterobacter, viii, 2, 21, 42
entrapment, 82
environment, 16, 46, 47, 55, 60, 72, 111
environmental conditions, 72
environmental factors, 21, 23
environmental impact, 64
environments, vii, 1, 72
enzymatic activity, 29
enzymatic hydrolysis, vii, ix, 27, 38, 47, 62, 64, 67, 71, 73, 75, 77, 78, 80, 81, 82, 84, 86, 87, 88, 89, 90, 91, 92, 93, 94, 96, 97, 98, 99, 100, 101, 103, 105, 106, 107, 108, 109, 110, 111, 112, 113, 114, 115, 116, 117, 118, 119, 120
enzyme(s), vii, viii, ix, 2, 3, 13, 14, 15, 18, 19, 20, 21, 22, 23, 24, 26, 28, 29, 31, 32, 33, 34, 36, 37, 41, 42, 44, 45, 46, 47, 48, 49, 50, 52, 53, 55, 56, 57, 58, 60, 62, 63, 64, 65, 66, 67, 69, 71, 72, 73, 74, 75, 76, 77, 78, 79, 80, 81, 82, 83, 84, 85, 86, 87, 88, 89, 90, 91, 92, 93, 94, 95, 96, 97, 98, 99, 100, 101, 103, 105, 107, 109, 110, 111, 113, 114, 116, 118, 119, 121
enzyme interaction, 99
enzyme-related factors, 81, 99
epidermis, 102
Escherichia, 5, 28, 38, 73, 85
ethanol, vii, viii, ix, 27, 35, 36, 47, 63, 65, 66, 67, 68, 69, 71, 72, 79, 80, 86, 87, 88, 89, 91, 92, 93, 94, 96, 100, 116, 118, 119, 120
Eubacterium, 31
eucalyptus, 23, 26, 117
eukaryotic cell, 85
exclusion, 111, 119
exopolysaccharides, 31
exposure, 105, 109, 110
extraction, 18, 29, 65, 103, 108, 113

F

Faecalibacterium, 31
families, 73, 74, 75
feedstock, ix, 27, 64, 67, 71, 79, 81, 83, 84, 86, 87, 88, 92, 93, 97, 99
fermentation, viii, 2, 13, 14, 15, 19, 21, 22, 24, 27, 29, 30, 32, 34, 36, 38, 39, 40, 41, 42, 44, 58, 65, 68, 71, 73, 74, 79, 80, 86, 87, 93, 94, 96
fertilizers, viii, 2, 4, 30, 39, 46
fiber, 27, 28, 30, 31, 65, 66, 78, 93, 98, 103, 104, 105, 111, 114, 120, 122
filtration, 29, 65
financial support, 57
first generation, 80
flavor, 29, 65
flexibility, 73
fluid, 15, 16, 35, 59
food, vii, viii, ix, 2, 27, 29, 46, 64, 66, 95, 121
formation, vii, 1, 9, 39, 43, 61, 65, 68, 104, 107, 114, 122
fruits, 12, 29, 65
functionalization, 122
fungi, viii, 2, 10, 11, 12, 13, 18, 19, 24, 26, 27, 29, 30, 32, 34, 35, 36, 38, 40, 41, 45, 46, 47, 48, 50, 52, 53, 56, 57, 60, 67, 72, 73, 76, 79, 85, 89, 111, 121

fungus, 19, 26, 36, 74
Fusarium, viii, 2, 17, 19, 25, 26, 34, 39, 72, 73
fusion, 94, 116

G

gene expression, 56
genetic engineering, 55, 113
genomics, 61, 86, 90
genus, 39, 40, 72
genus Streptomyces, 39
Geobacillus, viii, 2, 14, 16
germination, 30, 65
Gliomastix, 17
glucose, viii, ix, 2, 4, 9, 10, 12, 14, 23, 31, 46, 47, 48, 49, 50, 55, 56, 75, 77, 78, 80, 81, 82, 83, 87, 95, 98, 99, 100, 101, 103, 106, 107, 108, 109, 111, 113
glucosidases, 20, 49, 75, 83
glycerol, 5, 8, 9
glycoside, vii, ix, 1, 73, 89, 95
glycosylation, 85
grass, 16, 26, 102, 103, 106
growth, 5, 7, 11, 18, 22, 26, 30, 37, 50, 51, 53, 55, 60, 65, 70, 72, 102
growth rate, 5, 51, 70

H

health, vii, viii, 2, 31, 46, 55, 60
heat transfer, 80
heavy metals, 50, 53, 54, 57, 58, 61, 62
hemicellulose, 62, 67, 73, 81, 97, 101, 105, 114, 120
Hemicellulose(s), ix, 2, 62, 67, 73, 81, 83, 96, 97, 99, 100, 101, 105, 107, 110, 113, 114, 120
heterogeneity, 99, 101, 114
hot springs, 14, 15
human, viii, 2, 31, 38, 41, 46, 55
human health, viii, 2, 55
Humicola, 28, 29, 67, 72
hydrogen, x, 96, 98, 103, 104, 105, 115
hydrolysis, vii, ix, 3, 5, 6, 7, 11, 13, 15, 18, 27, 38, 39, 41, 47, 48, 50, 56, 62, 64, 67, 68, 71, 73, 75, 77, 78, 80, 81, 82, 83, 84, 86, 87, 88, 89, 90, 91, 92, 93, 94, 96, 97, 98, 99, 100, 101, 103, 105, 106, 107, 108, 109, 110, 111, 112, 113, 114, 115, 116, 117, 118, 119, 120, 121
hydrolysis kinetics, 108, 110
hydrolysis rate, 81, 87, 97, 99, 106, 107, 110
hydroxyl groups, 104

I

identification, 5, 11, 33, 34, 37, 38, 39, 44, 104
immune response, 31
immunogenicity, 61
improvements, 85
incubation period, 24, 48, 80, 82
incubation time, 23, 80, 92
India, 3, 6, 9, 13, 14, 15, 16, 17, 32, 33, 41, 42
inducer, 10, 34, 56
inducible enzyme, 26
induction, 40, 79
industrial sectors, ix, 95
industrial wastes, viii, 19, 24, 41, 46
industries, viii, 2, 3, 14, 24, 27, 28, 29, 30, 32, 35, 42, 64, 65, 66, 67, 94, 95
infection, 31
inhibition, x, 11, 49, 75, 82, 83, 86, 87, 94, 96, 98, 100, 101, 111
inoculation, 52, 54
inoculum, 10, 20, 21, 22, 23, 50
interaction process, 105
internode, 102, 103, 115
intestine, 15, 17, 44
isolation, vii, 1, 3, 4, 7, 8, 10, 12, 32, 33, 35, 38, 106, 119
Isoptericola, viii, 2, 15, 17, 33

K

kinetics, 108, 110
Klebsiella, viii, 2, 15, 16, 36

L

lignin, ix, 2, 53, 67, 78, 82, 83, 84, 86, 89, 91, 94, 96, 97, 100, 101, 102, 103, 105, 107, 108, 110, 112, 113, 114, 115, 117, 118, 119, 120
lignocellulosic biomass, 2, 27, 34, 38, 66, 75, 79, 83, 85, 88, 90, 96, 105, 115, 116, 117, 119, 121
lipids, 65
liquid phase, 109
liquids, 49, 73, 84, 86
Lithuania, 10, 17
local government, 17
Luo, 61, 119
Lysinibacillus, viii, 2, 15, 16
lysis, 30

M

machine learning, 42
macromolecules, 105, 106, 114
Malaysia, 37, 45, 57, 58
malt extract, 8, 9, 11, 21
manganese, 97
mangrove forests, 6
manipulation, 34
manufacturing, 61, 71
marine environment, 72
materials, vii, viii, ix, 1, 17, 26, 27, 40, 41, 45, 47, 49, 51, 52, 54, 64, 65, 66, 67, 71, 72, 80, 87, 88, 96, 98, 100, 104, 105, 107, 108, 117, 119, 120
mechanical properties, 103, 107
medical, 31, 39, 122
medium composition, 23
Melanocarpus, 37
Mesorhizobium, viii, 2, 15, 16, 21, 41
Metabolic, 85
metabolism, 79
metal ions, 20, 78
metals, 39, 50, 54, 57, 58, 61, 62
methodology, 35, 41
Mg^{2+}, 22, 62
microbial cells, 50, 52, 54
microbial community(ies), 31, 73
Micrococcus, viii, 2, 15, 16, 21, 23, 48, 52
microcrystalline cellulose, 3, 41, 92
microorganism, vii, viii, ix, 1, 3, 6, 14, 16, 22, 43, 46, 47, 48, 49, 50, 51, 52, 53, 54, 55, 56, 57, 58, 64, 71, 73, 87, 90, 114
microscopy, 115
Miscanthus, 120
mixing, 7, 87, 100
moisture, 24, 55, 79
molecular biology, 32, 93
molecular mass, 106
molecular weight, 15, 18, 50, 97
molecules, 27, 49, 66, 78, 103, 104, 105, 106, 111, 114
Morganella, viii, 2, 16
morphology, 5, 79, 103, 114
Mucor, viii, 2, 17, 19, 20, 25, 26, 27, 40
municipal solid waste, 24, 59
Myceliophthora, viii, 2, 17, 19

N

NaCl, 3, 4, 5, 6, 7, 8, 9, 10, 11, 13, 21, 23, 48
nanofibers, 103
nanomaterials, 103
National Academy of Sciences, 43
natural selection, 55
negative relation, 112
nitrogen, 15, 20, 22, 23, 24, 26, 50, 53, 58, 59, 111
nitrogen compounds, 53
NREL, 77
nutrient, 3, 7, 32, 43, 55

O

Ochrobactrum, viii, 2, 15, 17, 35
OECD, 64, 92
oil, 15, 17, 29, 32, 37, 38, 57, 58, 62, 89
oligomers, 87, 109
opportunities, 91
optimization, ix, 20, 23, 37, 38, 40, 64, 89
organic compounds, 49
organic matter, 17, 19
organism, viii, 45, 55, 56
oxygen, 49, 75, 79

P

Paenibacillus, viii, 2, 14, 15, 16, 21, 22, 28, 32, 38, 39, 53, 59
palm oil, 57, 58, 62
Paracoccus, 15, 16, 35
parenchyma, 100, 102
pathogens, 30, 46
Penicillium, viii, 2, 17, 18, 40, 72, 91
peptide, 75, 77
pH, 4, 5, 6, 8, 9, 12, 13, 14, 15, 18, 19, 20, 21, 22, 23, 24, 25, 26, 28, 35, 47, 48, 50, 52, 54, 55, 66, 67, 72, 79, 82, 101, 108
pharmaceutical, 31, 50, 55, 57, 58, 92, 122
phosphate, 4, 48
phosphorus, 42
phylogenetic tree, 59
physical properties, 97, 110
physical structure, 103, 108, 120
physicochemical properties, x, 96, 98, 99, 100, 103, 106, 108, 115
Pichia, 29, 85
plant cell wall(s), ix, 89, 96, 98, 100, 102, 108, 114
plant growth, 30, 65
plants, 11, 46, 49, 54, 58, 67, 68, 72, 103, 104, 112, 116
polymerization, 75, 81, 97, 99, 101, 102, 106, 114, 116, 118, 119
polymers, x, 2, 72, 86, 96
polysaccharide, 43, 47, 67, 69, 78, 81, 83, 88, 89, 92, 97, 98, 108, 117, 118
polythene, 9
population, 51, 61
porosity, 78, 81, 98, 101, 102, 111
potassium, 5
producers, vii, ix, 2, 3, 13, 17, 32, 64, 70, 71, 85
production costs, 84
project, 57
prosthetic materials, 41
protection, 114
proteins, 30, 76, 85, 86, 114
Providencia, viii, 2, 16
Pseudomonas, viii, 2, 15, 16, 21, 22, 31, 37, 44, 47, 72
Pseudomonas aeruginosa, 31, 37
pulp, vii, ix, 4, 6, 10, 24, 25, 27, 34, 38, 40, 42, 65, 66, 67, 95, 121
purification, 18, 32, 33, 34, 36, 61, 73, 80

R

reactions, 49, 72, 74, 107, 116
reactivity, 75, 89, 97, 104, 107, 115, 117
recalcitrance, ix, 63, 82, 83, 84, 85, 97, 98, 99, 100, 101, 102, 103, 104, 106, 108, 112, 114, 115, 116, 120
recognition, 91, 100
recycling, 3, 40, 78, 86
regenerated cellulose, 104
regulatory system, 55
reinforcing fibers, 103
repression, vii, viii, 46, 55, 56, 60, 62, 79
requirements, ix, 64, 65, 67, 79, 83, 84, 97, 100
residues, 2, 19, 24, 27, 29, 33, 34, 38, 55, 67, 68, 74, 75, 121
resistance, 59, 61, 67, 73, 83, 97, 101, 106, 113

resources, 47, 55, 67
response, 31, 35, 41, 98, 110, 115
restoration, 43
rheology, 122
Rhizopus, 17, 33
riboflavin, 8
root, 9, 30, 65, 94, 119

S

saccharification, ix, 27, 29, 33, 41, 43, 64, 67, 69, 71, 86, 87, 91, 94, 97, 99, 117, 118, 121
sawdust, 24, 27, 41, 53, 92
scanning calorimetry, 111
Schizophyllum, viii, 2, 25, 26, 38, 72
scientific knowledge, 32
sclerenchyma, 99, 100, 102
second generation, viii, 63, 66, 67, 68, 69, 71, 72, 96
secretion, 10, 73, 85, 89
Serratia, 31
Shigella, viii, 2, 14, 16, 43
SIGMA, 11
sodium, 4, 6, 11, 12, 13, 20
softwoods, 90, 117, 118
solid phase, 109
solid state, viii, 2, 19, 24, 34, 40, 44, 60, 79
solids loading, 71, 87, 94, 100
solution, 4, 6, 8, 9, 10, 11, 13, 48, 49
solvents, 64, 65, 84, 103, 110
South Africa, 63, 95, 117
species, 2, 11, 13, 14, 18, 40, 42, 47, 48, 52, 54, 60, 61, 102, 104, 119
specific surface, 110, 111
Sphingomonas, viii, 2, 21, 23, 39
Sporotrichum, viii, 2, 17, 19, 29, 33
stability, x, 19, 28, 29, 65, 72, 96, 101, 114
stabilization, 29, 46, 57
Stachybotrys, viii, 2, 17, 18
starch, 4, 6, 7, 8, 9, 10, 23
Streptomyce(s), viii, 2, 16, 17, 21, 23, 30, 35, 36, 37, 39, 41, 42, 50, 72
Streptomycetes, 38
structural carbohydrate, 114
structural characteristics, 80
structural heterogeneity, 101
structural variation, 119
structure, 61, 66, 74, 75, 76, 77, 82, 90, 93, 97, 99, 100, 102, 103, 104, 109, 110, 112, 113, 114, 117, 118, 120
substrate, vii, ix, 7, 8, 9, 10, 13, 15, 19, 20, 21, 22, 23, 24, 26, 28, 29, 31, 32, 34, 38, 40, 41, 43, 47, 48, 49, 51, 53, 54, 55, 56, 64, 67, 73, 74, 75, 78, 79, 80, 81, 82, 84, 85, 86, 88, 89, 91, 92, 93, 94, 96, 97, 98, 99, 100, 101, 107, 109, 111, 112, 114, 115, 116, 117, 118
substrate-related factors, 81, 99
sugar beet, 24, 40
sugar industry, 14, 24
sugarcane, viii, 2, 4, 6, 11, 13, 15, 19, 26, 27, 28, 35, 89, 94, 102, 103, 110, 115, 118, 119
surface area, 78, 100, 101, 109, 110, 111
surface modification, 122
survival, 55, 56
susceptibility, 86, 97, 107
swelling, 78, 103, 110
synergistic effect, viii, 46, 91
synthesis, 34, 44, 61

T

Taiwan, 4, 16
Talaromyces, viii, 2, 17, 19, 36, 72
techniques, vii, 1, 6, 20, 119, 122
technological advances, 122
technology(ies), ix, viii, 46, 47, 49, 55, 56, 58, 59, 62, 63, 67, 70, 80, 83, 84, 88, 90, 93, 96, 97, 99, 114, 115, 117
temperature, x, 5, 7, 18, 20, 22, 23, 24, 28, 35, 36, 47, 50, 52, 66, 67, 72, 82, 87, 96, 103, 120
tempo, 119
tensile strength, 104
Thailand, 1, 6, 15, 16, 17, 34

thermal treatment, 57
Thermomonospora, 28, 33, 72
Thermomucor, viii, 2, 19
Thermomyces, 19
thermostability, 67, 83, 86, 88
tissue, 99, 100, 101, 102, 103, 114
treatment, viii, 10, 38, 45, 46, 47, 49, 50, 51, 52, 53, 54, 55, 56, 57, 58, 60, 61, 62, 66, 67, 113, 117, 122
Trichoderma, vii, viii, ix, 2, 17, 18, 24, 25, 27, 28, 29, 36, 39, 40, 41, 43, 52, 60, 64, 67, 71, 72, 73, 74, 90, 91, 92, 93, 118

U

ultrastructure, 116
United States(USA), 43, 59, 68, 70, 119
urea, 12, 24, 53

V

valorization, ix, 96
varieties, 89, 98, 102, 115
vascular bundle, 99, 102
vegetables, 11, 12, 29, 65
versatility, 93, 118
vessels, 100, 102
Vibrio, viii, 2, 16
Vietnam, 7, 18, 34
viscosity, 11, 29, 65, 72, 87
vitamin B12, 4, 8

W

waste, ix, 7, 10, 12, 15, 16, 17, 18, 21, 22, 24, 25, 28, 29, 30, 32, 33, 34, 37, 38, 39, 40, 41, 42, 44, 59, 60, 64, 66, 93, 96
waste treatment, 60
waste water, 10, 12, 17, 42, 48, 49, 50, 54, 55, 58, 60, 61, 62, 122
water, 4, 6, 7, 9, 10, 12, 13, 14, 17, 64, 73, 94, 103, 110, 111, 120
wood species, 104
worldwide, viii, 2

X

Xanthomonas, viii, 2, 15, 16
xylem, 102

Y

yeast, 4, 8, 9, 10, 12, 15, 20, 22, 23, 24, 27, 29, 53, 85, 94
Yemen, 45, 57
Yersinia, viii, 2, 16
yield, 20, 22, 23, 24, 27, 29, 66, 79, 80, 81, 87, 89, 98, 99, 100, 101, 103, 107, 108, 109